WATER MANAGEMENT MODELS
A Guide to Software

WATER MANAGEMENT MODELS

A Guide to Software

RALPH A. WURBS
Civil Engineering Department
Texas A&M University

PRENTICE HALL PTR
Englewood Cliffs, New Jersey 07632

Library of Congress Cataloging-in-Publication Data

Wurbs, Ralph Allen.
    Water management models : a guide to software / Ralph A. Wurbs.
      p. cm.
    Includes bibliographical references and index.
    ISBN 0-13-161621-8
    1. Water-supply—Management—Computer programs. I. Title.
TD353.W87 1995
628.1′0285′5369—dc20   94-23463
                                                    CIP

Cover design: Design Source
Acquisitions editor: Karen Gettman
Manufacturing manager: Alexis Heydt
Production management: Superscript Editorial Production Services

© 1995 by Prentice Hall PTR
Prentice-Hall, Inc.
A Simon & Schuster Company
Englewood Cliffs, New Jersey 07632

The publisher offers discounts on this book when ordered in bulk quantities. For more information, contact: Corporate Sales Department, Prentice Hall PTR, 113 Sylvan Avenue, Englewood Cliffs, NJ 07632; Phone: 800-382-3419, Fax: 201-592-2249, E-mail: dan_rush@prenhall.com.

Printed in the United States of America
10 9 8 7 6 5 4 3 2 1

ISBN 0-13-161621-8

Prentice-Hall International (UK) Limited, *London*
Prentice-Hall of Australia Pty. Limited, *Sydney*
Prentice-Hall Canada Inc., *Toronto*
Prentice-Hall Hispanoamericana, S.A., *Mexico*
Prentice-Hall of India Private Limited, *New Delhi*
Prentice-Hall of Japan, Inc., *Tokyo*
Simon & Schuster Asia Pte. Ltd., *Singapore*
Editora Prentice-Hall do Brasil, Lida., *Rio de Janeiro*

CONTENTS

PREFACE   ix

1   INTRODUCTION   1

Role of Models in Water Management   1
Computer Modeling   3
Generalized Operational Water Management Models   5
Computer Hardware   7
Computer Software   10
Modeling Systems   13
References   15

**2   MODEL DEVELOPMENT AND DISTRIBUTION
      ORGANIZATIONS**   17

The Model Development Community   18
International Perspective   18
Federal Agencies   20
U.S. Army Corps of Engineers   20
U.S. Geological Survey   23
National Weather Service   24
Soil Conservation Service   25
Bureau of Reclamation   25
Environmental Protection Agency   25
International Ground Water Modeling Center   28

National Technical Information Service   29
McTrans Center for Microcomputers in Transportation   29
References   32

3   GENERAL-PURPOSE SOFTWARE   33

Introduction   33
Spreadsheet/Graphics/Database Software   35
Object-Oriented System Simulation Software   37
Comparison of Alternative Model Building Approaches   39
Expert Systems   42
Equation Solvers and Mathematical Modeling Environments   45
Mathematical Programming (Optimization) Models   46
Statistical Analysis Programs   48
Data Management Systems   50
Geographic Information Systems   54
Computer-Aided Drafting and Design   59
Graphics Programs   60
References   62

**4   DEMAND FORECASTING AND BALANCING SUPPLY
     WITH DEMAND   68**

Introduction   68
Municipal and Industrial Water Use Forecasting   68
Agricultural Water Requirements   76
Water Evaluation and Planning System   76
References   78

5   WATER DISTRIBUTION SYSTEM MODELS   81

Introduction   81
Review of Available Models   84
Models Included in the Model Inventory Appendix   84
References   86

6   GROUND-WATER MODELS   88

Introduction   88
Review of Available Models   92
Comparison of Models Included in the Model Inventory Appendix   102
References   103

7   WATERSHED RUNOFF MODELS   107

Introduction   107

Review of Available Models   110
Comparison of Models Included in the Model Inventory Appendix   120
References   122

8 STREAM HYDRAULICS MODELS   125

Introduction   125
Review of Available Models   128
Comparison of Models Included in the Model Inventory Appendix   136
References   137

9 RIVER AND RESERVOIR WATER QUALITY MODELS   139

Introduction   139
Review of Available Models   142
Comparison of Models Included in the Model Inventory Appendix   148
References   149

10 RESERVOIR/RIVER SYSTEM OPERATION MODELS   153

Introduction   153
Review of Available Models   156
Comparison of Models Included in the Model Inventory Appendix   176
References   179

MODEL INVENTORY APPENDIX   183

Demand Forecasting and Balancing Supply with Demand Models
    (Chapter 4)   183
    (IWR-MAIN) Water Use Forecasting System   183
    (WEAP) Water Evaluation and Planning System   184
Water Distribution System Analysis Models (Chapter 5)   185
    (KYPIPE2) University of Kentucky Pipe Network Model   185
    (WADISO) Water Distribution Simulation and Optimization
    Model   187
Ground-Water Models (Chapter 6)   188
    (MODFLOW) Modular Three-Dimensional Finite-Difference
    Ground-Water Flow Model   188
    (PLASM) Prickett-Lonnquist Aquifer Simulation Model   189
    (RANDOM WALK) Random Walk Solute Transport Model   190
    (MOC) Method of Characteristics Two-Dimensional Solute
    Transport Model   191
    (WHPA) Wellhead Protection Area Model   192
    (SUTRA) Saturated-Unsaturated Transport Model   193
    (SWIFT-II) Sandia Waste Isolation, Flow and Transport Model   194

Watershed Runoff Models (Chapter 7)   195
    (HEC-1) Flood Hydrograph Package   195
    (TR-20) Computer Program for Project Hydrology   197
    A&M Watershed Model   198
    (SSARR) Streamflow Synthesis and Reservoir Regulation199
    (SWMM) Stormwater Management Model   200
    (HSPF) Hydrologic Simulation Program—Fortran   201
    (SWRRB-WQ) Simulator for Water Resources in Rural Basins—
        Water Quality   202
Stream Hydraulics Models (Chapter 8)
    (HEC-2) Water Surface Profiles   203
    (WSPRO) Water Surface Profiles   204
    (FLDWAV) Flood Wave Model   205
    (UNET) One-Dimensional Unsteady Flow through a Full Network
        of Open Channels   207
    (FESWMS-2DH) Finite Element Surface-Water Modeling System:
        Two-Dimensional Flow in a Horizontal Plane   208
    (HEC-6) Scour and Deposition in Rivers and Reservoirs   209
River and Reservoir Water Quality Models (Chapter 9)
    (QUAL2E) Enhanced Stream Water Quality Model   210
    (WASP) Water Quality Analysis Simulation Program   211
    (CE-QUAL-RIV1) A Dynamic One-Dimensional Water Quality
        Model for Streams   212
    (CE-QUAL-R1) A Numerical One-Dimensional Model of
        Reservoir Water Quality   213
    (CE-QUAL-W2) A Numerical Two-Dimensional, Laterally
        Averaged Model of Hydrodynamics and Water Quality   215
    (HEC-5Q) Simulation of Flood Control and Conservation
        Systems (Water Quality Version)   216
    (WQRRS) Water Quality for River-Reservoir Systems   217
Reservoir/River System Operation Models (Chapter 10)
    (HEC-5) Simulation of Flood Control and Conservation Systems   218
    (IRIS) Interactive River System Simulation Model   219
    (TAMUWRAP) Water Rights Analysis Package   220
    (MODSIM) River Basin Network Simulation Model   221
    (HEC-PRM) Prescriptive Reservoir Model   222
    (RSS) River Simulation System   224
    (CALIDAD) Object-Oriented River Basin Modeling Framework   225
Software Index   **227**
Index   **237**

PREFACE

An extremely large number of water management models that provide a broad range of analysis capabilities have been reported in the published literature. Many other computer programs have been developed and applied successfully by water agencies and firms without being published. Commercial software products, which are used extensively in business, education, science, engineering, and other professional fields, also play important roles in environmental and water resources planning and management. The number and complexity of software packages can be overwhelming for anyone attempting to sort through the maze of models and understand which ones might be most useful for a particular application. The purpose of this book is to assist water managers, planners, engineers, scientists, and educators in locating the optimal computer programs for their particular needs.

This book deals with generalized operational water management models, as defined in Chapter 1. The majority of the models cited are public domain. Federal agencies and other entities that develop and distribute water management computer programs are discussed in Chapter 2. Chapter 3 covers general-purpose software packages that are applicable to several or all of the water management model categories of Chapters 4 through 10, either as complete models or as pre- and postprocessor programs. The software cited in Chapter 3 includes popular commercial programs marketed by the software industry as well as models developed within the water management community. The several hundred models noted in this book are a relatively small subset of all the water management models reported in the literature. Citing of particular computer programs does not imply that other excellent generalized operational water management models are not available.

Chapters 4 through 10, respectively, are devoted to each of the following categories of water management models:

Chapter 4—models for demand forecasting and balancing supply with demand
Chapter 5—water distribution system models
Chapter 6—ground-water models
Chapter 7—watershed runoff models
Chapter 8—stream hydraulics models
Chapter 9—river and reservoir water quality models
Chapter 10—reservoir/river system operation models

Of course, computer models are used in many other areas of water management as well. However, the selected categories account for a major portion of the more common frequent applications. Computer programs in each of these major categories are applied widely by water resources and environmental professionals. Each of the seven chapters include (1) an introductory overview of modeling applications, processes being modeled, pertinent literature, and model categorization; (2) a review of available models; and (3) a comparison of programs selected for inclusion in the Model Inventory presented as an appendix.

The Model Inventory Appendix includes a collection of 38 generalized water management models representing the seven categories of Chapters 4 through 10. Thus, a general overview of many models is combined with a focus on a selected few. An attempt was made to select several models covering a broad range of capabilities in each category. Preference was given to models with a proven record of successful application. A few of the models are more or less accepted standards in the water management community for certain types of computations; however, several of the selected models are relatively new and have not yet been applied widely. Inclusion of only selected models in the Inventory Appendix does not mean other excellent models are not also available. Practitioners should consider the entire spectrum of modeling methodologies in formulating a modeling and analysis approach for their specific applications. However, the models included in the Appendix provide an excellent starting point for investigating available capabilities. These highlighted models are readily available generalized software packages that provide a broad range of analysis capabilities and are designed for practical application.

ACKNOWLEDGMENTS

This book was written under the auspices of the National Study of Water Management During Drought conducted by the U.S. Army Corps of Engineers Institute for

Water Resources. The author worked for the Institute for Water Resources (IWR) under an Intergovernmental Personnel Act Agreement while preparing the book. William J. Wrick, Eugene Z. Stakhiv, and Germaine A. Hofbauer, of IWR, conceived the concept of a water management software guide and guided the project to completion. Their superb guidance and support are gratefully acknowledged. Numerous model developers and users contributed information essential to the preparation of the book. The generous sharing of time and expertise of all the individuals contacted during the course of the project is greatly appreciated. Suggestions and comments provided by the several individuals who reviewed the draft manuscript were also helpful. Thanks are extended to Darryl W. Davis, Richard J. Hayes, and others at the Hydrologic Engineering Center; Mark S. Dortch, Waterways Experiment Station; Paul K. M. van der Heijde, International Ground Water Modeling Center; and Anil R. Yerramreddy, Texas Natural Resource Conservation Commission, for their reviews. Of course, the responsibility for all errors and omissions remains with the author. Mention of trade names or commercial products does not imply their endorsement by the United States government.

Ralph A. Wurbs

1

INTRODUCTION

A tremendous amount of work has been accomplished during the past three decades in developing computer models for use in water resources planning and management. Powerful generalized software packages are playing an increasingly important role in all aspects of water management. With recent advances in computer technology, essentially everyone working in the water resources and environmental field now has convenient access to desktop computers that provide all the hardware capabilities needed to execute a mighty arsenal of available models. Computer modeling is no longer limited solely to organizations that can afford expensive mainframe computers. The purpose of this book is to make available computer programs more accessible to water management professionals. Information is provided to facilitate the identification, selection, and acquisition of software packages for a broad spectrum of water management modeling applications.

ROLE OF MODELS IN WATER MANAGEMENT

Water management involves the development, control, protection, regulation, and beneficial use of surface (rivers and reservoirs) and ground-water resources. Services provided by the water management community include water supply for agricultural, industrial, and municipal uses; wastewater collection and treatment; protection and enhancement of environmental resources; pollution prevention; recreation; navigation; hydroelectric power generation; stormwater drainage; erosion and sedimentation control; and controlling flood waters and reducing damages due to flooding. Water resources planning and management activities include policy formulation; national, regional, and local resource assessments; regulatory and permitting functions; formulation and implementation of resource management strategies; planning, design, construction, maintenance, and operation of structures and facilities; scientific and engineering research; and education and training.

Computer models play important roles in all aspects of water management. There is a great diversity of ways in which they are applied. During the construction era of the 1950s through the 1970s, model development was oriented toward the water management emphasis on planning and design of constructed facilities, and these types of applications continue to be important. However, water resources management and associated modeling applications shifted during the 1970s and 1980s to a major focus on resource management and operation of existing facilities. Protection of water quality and environmental resources has become a driving concern in recent years.

The scale and complexity of problems analyzed with models vary greatly. For example, one or several software packages may be applied in the design of an urban stormwater detention basin with a watershed area of less than 1 square mile. Other models, or perhaps even the same models, may be used to optimize the operations of a major multiple-reservoir, multiple-purpose (flood control, water supply, hydropower, recreation) system regulating the water resources of a river basin with a drainage area of many thousand square miles. At the local level, models may be used both by a consulting firm to prepare a permit application for expansion of the wastewater treatment plant of a small municipality and by a state regulatory agency to evaluate the permit application. Likewise, on a larger geographic scale, models may be used to formulate and evaluate plans for solving regional or basinwide nonpoint source pollution problems. Models may be used to predict drawdown conditions to be expected as a small but growing community increases its ground-water pumpage or alternatively to assess the ground-water resources available from a large regional aquifer supplying water users located in several states.

From the perspective of the scientist and researcher, the role of mathematical models is to contribute to a better understanding of real-world processes. From the perspective of the water manager, the role of mathematical models is to provide quantitative information to support decision-making activities. Models help both individual water managers and the water management community to make better decisions. Models do not relieve people of the burden of making difficult decisions. They simply provide some additional information to consider. Models strengthen the knowledge base which supports decision-making processes.

In 1982, the Office of Technology Assessment of the U.S. Congress completed a comprehensive assessment of the use of mathematical models in water resources planning and management. Numerous experts from agencies, universities, and consulting firms participated in the study. The findings were broad in scope and included the following observations (Office of Technology Assessment 1982).

- Models capable of analyzing many pressing water resource issues are available and have significant potential for increasing the accuracy and effectiveness of information available to managers, decision makers, and scientists. Models have significantly expanded the nation's ability to understand and manage its water resources. Models have the potential to provide even greater benefits for water resource decision making in the future.
- Although the federal government spends about $50 million annually on water-related mathematical models, such tools are instrumental in planning billions of dollars of annual water resource investments and managing hundreds of billions of dollars of existing facilities.
- Models are used in essentially every area of water resources planning and management.
- Water resource models vary greatly in their capabilities and limitations and must be selected and used carefully by knowledgable professionals.
- Most states depend on the federal government to provide suitable models.
- Virtually all federal modeling activities are managed on an agency-by-agency basis. No entity within the federal government is specifically charged with providing information to potential users, federal or nonfederal, about the governmentwide availability of water resource models.

Since 1982, when the Office of Technology Assessment report was published, the use of water management models has exploded with the advent of the microcomputer. The model-user community has grown dramatically, particularly in regard to local public agencies, private consulting firms, and other nonfederal users. Most of the water management models cited throughout this book include user-friendly executable (ready-to-run) versions for desktop computers. Essentially everyone in the water management community now has convenient access to the computer hardware needed to run the available software.

COMPUTER MODELING

Development and application of water management models requires a thorough understanding of

- The role of the models in the overall water resources management decision-making process and the questions to be answered by the modeling exercises
- The real-world processes being modeled and the capabilities and limitations of methods for representing these processes with mathematical equations
- Computational techniques for solving the equations
- Data availability and limitations
- Model calibration and verification techniques
- The availability of computer software and hardware and the skills required to use these tools
- The communication capabilities required to assure that model development and application are responsive to water resources planning and management needs and that model results are effectively incorporated into decision-making processes

A team of several or many professionals may be required to assemble the necessary knowledge and capabilities to conduct a modeling study successfully. Water managers and planners play key roles in establishing the objectives to be accomplished and questions to be answered by the modeling exercise and incorporating model results into decision-making processes. Engineers and scientists from various disciplines provide the expertise required to (1) understand the real-world processes of concern, (2) capture the essence of these processes with mathematical expressions and data, (3) develop values for model parameters, and (4) efficiently solve the resulting systems of equations. Expertise is also required to select, acquire, and apply available software, develop computer programs, and use computer systems.

A variety of approaches and software tools are available for building models. A key question in developing a model for a particular study is whether to

- Construct a model within a modeling software environment such as those provided by the spreadsheet programs, object-oriented system simulation packages, equation solvers, and other commercially marketed software discussed in Chapter 3.
- Use an existing generalized water management model such as those cited in Chapters 4 through 10.
- Modify an existing model.
- Develop a new program using the traditional FORTRAN, BASIC, C, or some other programming language.
- Develop a new program using an object-oriented programming language such as C++.
- Adopt some combination or variation of the aforementioned.

The scope of this book is limited to available generalized water management models and off-the-shelf commercial software. Development of new computer programs, written in the various versions of FORTRAN and/or C or other commercially available language software packages, is another important modeling activity which is outside the scope of this book. Improved and expanded modeling approaches and methods continue to be implemented by writing new computer programs. Specific analysis needs may warrant coding new software specifically for a particular water resources system and/or a particular type of analysis. Many engineers and scientists naturally prefer the flexibility of working with programs they have coded themselves. Computer programs can be written from scratch fairly easily for relatively simple models. However, for complex models, formulating algorithms, devising data management schemes, writing and debugging code, and testing new programs are extremely time consuming and expensive. Attention and resources devoted to the nuts and bolts of computer programming can distract from focusing on the actual water management concern. Application of existing, readily available generalized software is often either the optimal use of available funding and time resources or essentially the only feasible way to conduct a study.

GENERALIZED OPERATIONAL WATER MANAGEMENT MODELS

This book deals with generalized operational water management models. Definition of these terms is appropriate in this introductory chapter. The term *model* can be defined as any simplified representation of a real-world system. Interest in this book is limited to representing water-related systems with mathematical formulations which are solved using a computer. This book focuses on computer simulation of natural and human-altered water resources systems.

Generalized means that the computer model is designed for application to a range of problems that deal with systems of various configurations and locations, rather than being developed to address a particular problem at a specific site. For example, a generalized model might be applicable to various ground-water systems or to essentially any river/reservoir system, rather than one particular system. The term *model* is used herein primarily to mean a generalized software package. A water management model for a particular problem may consist of a generalized software package combined with input data developed for the specific application.

Operational means that the model is reasonably well documented and tested and is designed to be used by professional practitioners other than the original model developers. Generalized models should be convenient to obtain, understand, and use and should work correctly, completely, and efficiently. Documentation, user support, and user friendliness of the software are key factors in selecting a model. The extent to which a model has been tested and applied in actual studies is also an important consideration.

Most of the thousands of water management computer models reported in the literature are not generalized and operational. Many models were developed to support a particular study, and when the study was completed, the model was shelved. Many models used to support operation of existing facilities continue to be applied successfully for a particular system but are not generalized for application to other systems. Other models contribute significantly to educating the developers but never reach the stage of being operational. Examples include many models developed for university graduate student theses and dissertations but include many models developed by practicing professionals as well.

Generalized software systems are, of course, used to manage data, perform computations to simulate real-world processes, and display results for various analysis applications. Computer models also play an important role in documenting and transferring knowledge. Textbooks, engineering manuals, and other written library references have traditionally served to record knowledge regarding real-world processes and analysis methods for studying these processes. Generalized water management modeling packages, including computer programs and documentation, also serve as textbooks to organize, record, and pass on state-of-the-art knowledge.

A model for a particular study of a specific water resources system consists of user-developed input data files combined with one or more generalized software packages. Regardless of the sophistication of a generalized computer program, the quality of the modeling results for a particular application can be no better than the input data for that application. Parameter values are required for the governing equations representing the processes being modeled. A variety of other input data are also required. Development of the often voluminous input data is a key aspect of modeling.

User-supplied values are required for the parameters of the equations embedded in generalized models. Calibration and verification are crucial to modeling studies and typically require significant expertise and effort. Some generalized modeling packages provide features to facilitate calibration. For example, certain pipe network analysis (Chapter 5) and stream hydraulics (Chapter 8) packages include capabilities for computing values for head loss coefficients which best reproduce inputted measured hydraulic grades or water surface profiles. Watershed models (Chapter 7) may include features that partially automate determination of parameter values that best reproduce the inputted measured runoff hydrographs resulting from each of the particular rainfall events used for calibration. Groundwater models (Chapter 6) may solve the inverse problem of computing parameters that characterize an aquifer based on measured drawdowns resulting from well pumping.

A word of caution is warranted regarding the use of generalized water management models. It is hoped that this book will contribute to increasing awareness within the water management community of the arsenal of powerful tools which are readily available. This will be a significant contribution to water management. However, the danger always exists of providing novices with a weapon with which to shoot themselves in the foot. Easy access to computer software does not diminish the necessity for high levels of technical knowledge and expertise. The user of off-the-shelf software must have a thorough understanding of the computations performed by the model and the capabilities and limitations of the model in representing real-world processes. Models must be applied carefully and meticulously with professional judgment and good common sense. Although the effectiveness and efficiency of a modeling study can be greatly enhanced by exploiting the capabilities provided by readily available software, modeling requires significant time and effort as well as expertise.

Generalized operational water management models evolve as new versions become available periodically. Software development and maintenance are expensive, and funding is a key consideration in providing ongoing model support. Feedback from users can contribute greatly to ongoing model improvements. Ideally, model revisions and expansions should reflect the cumulative experience of multiple users and developers. Training and technical support are important for new users. Sharing of experiences and lessons learned is also important for experienced model users. An institutional capability is needed to continually maintain, update, and improve a model and its documentation and to support users in applying the model. Available computer programs must be disseminated to the user community to reap the potential benefits of research and development efforts. Institutional aspects of model development and availability are addressed in Chapter 2.

A number of generalized operational models applied in several major areas of environmental and water resources planning and management are noted in Chapters 4 through 10. Several of these models are also included in the Model Inventory Appendix. The general-purpose software packages discussed in Chapter 3 are applicable to several or all of the categories of water management modeling applications addressed in Chapters 4 through 10. Chapter 3 covers popular commercial software products as well as public domain computer programs developed by water resources engineers and scientists. The intent of this book is to outline categories of software useful in water management and to highlight several representative programs in each category. There are many other excellent generalized operational water management models which are not mentioned in this book.

COMPUTER HARDWARE

The full range of computer sizes (microcomputers, workstations, minicomputers, mainframes, and supercomputers) have been and continue to be used in water resources planning and management. Hopper and Mandell (1990) outline the fundamentals of computer systems. Henle and Kuvshinoff (1992) focus specifically on microcomputers. Early computers were large-scale mainframe systems. The name *mainframe* comes from the fact that the processor of these computers consists of a series of circuit boards mounted within a frame structure. Since the mid-1950s, the mainframe computer field has been dominated by IBM. Competing companies include Burroughs, Control Data, Honeywell, NCR, and UNIVAC. The Digital Equipment Company (DEC) has played a key role in manufacturing and marketing minicomputers. The minicomputer got its name in comparison with the large-scale computers that dominated the market when it was introduced in the mid-1960s. During that era, all computers were large and expensive tools for only the largest business, scientific, and government organizations. Less expensive minicomputers were introduced to provide computer capabilities for medium-sized organizations. Supercomputers are powerful mainframes that represent the upper extreme in size, computational power, and cost. Cray Research has been a pioneering company in developing and marketing supercomputers. Other companies in the supercomputer business include Control Data Corporation and NEC.

The microcomputer gets its name from the fact that it uses a microprocessor on a single chip as its central processing unit. Microcomputers are also called desktop computers and personal computers. Apple Computer, Inc. introduced the first commercially successful microcomputer, the Apple II, in 1977. Apple first marketed the popular Macintosh computer in 1984. The Power Macintosh, based on the much faster PowerPC microprocessor, was introduced by Apple in 1994. IBM introduced its 8088-based PC (personal computer) in 1981, followed in 1984 by the PC AT (personal computer, advanced technology) based on the 80286 microprocessor. IBM early adopted a version of Microsoft's Disk Operating System (MS-DOS) for its operating system software. Because of standards adopted by IBM, many other companies have, since the mid-1980s, marketed IBM-compatible microcomputers, which also operate under MS-DOS and run the same applications software. The IBM-compatible 80386-based systems that became popular in 1989 were subsequently followed by computers incorporating the 80486 microprocessor. The much faster Pentium microprocessor was introduced in 1993. Today's microcomputers provide computational and data management capabilities that greatly exceed those provided by the largest of mainframe computers 20 years ago. The $2,000 price of an IBM-compatible desktop computer based on the 80486 microprocessor in 1994 is a small fraction of the multi-million-dollar expense required to own a mainframe computer 20 years ago.

In the mid-1980s, workstations were introduced that blur the distinctions between desktop computers and larger scale minicomputer and mainframe computer systems. Workstations are supermicrocomputers that use microprocessors as their central processing units but typically include coprocessors that expand data-handling capabilities. Some coprocessors add to mathematical capabilities, and others specialize in the control of databases. Workstation systems provide excellent graphical display as well as computational capabilities. Companies that market workstations include Sun Microsystems, Apollo, NeXT, NCR, and AT&T. Unix is a popular operating system often used on workstations. Various versions of Unix are used on desktop, mini, and mainframe computers as well as workstations. MS-DOS is also sometimes used on workstations. With recent advances in microcomputer hardware and software, the distinctions between desktop and workstation computer systems are diminishing. Advanced IBM-compatible 80486-based computers, running under either Unix or MS-DOS, are often considered to be workstations.

In many organizations, desktop computers are interconnected in local area networks (LANs), which may also include access to one or more workstations, minicomputers, or mainframe computers. The LAN allows numerous personnel to have individual microcomputers on their desks but share peripheral hardware, such as printers, as well as software and databases stored on a common disk.

Many of the fundamentals of representing real-world water resources systems by mathematical formulations and solving the equations using computers were developed during the 1960s and 1970s. A majority of the water management models cited in this book were either developed during the period from the mid-1960s through the 1970s or were developed later based on updating and expanding existing models that were developed during that era. Models tend to evolve over time through various versions or from earlier models. Thus, many of the presently available models were originally developed and applied on mainframes and minicomputers.

A major focus in water management model development during the mid-1980s was on converting models, which were originally developed for mainframes and minicomputers, to run on desktop computers. IBM-compatible desktop computers, operating under MS-DOS, became the accepted standard microcomputer system for water management applications as well as for many other fields. Programs began to be distributed on diskette as compiled executable versions for IBM PC-compatible desktop computers operating under MS-DOS. During the later 1980s, a major emphasis was on taking advantage of microprocessor technology to develop user-friendly interactive human-machine interfacing; improve data management capabilities; utilize graphics intensively; and enhance interpretation and communication of model results.

Microcomputers have had a great impact on water management modeling since the early 1980s, particularly in regard to the following:

- Creation of an entire new market of popular software for business, science, and engineering applications, including water management

- Expansion of the number of firms and individuals using water management computer programs
- Reduction of the difficulties with transporting software between computer systems
- Development of a focus on user-oriented software

The significantly more expensive workstations, typically operating under Unix, have established a more limited but yet significant role in water management in recent years. Mainframe and minicomputers also continue to be used widely in water resources planning and management. The large-scale systems have significant advantages for multiple users in large organizations. Some computer models either require or significantly benefit from the greater speed and storage capabilities of the larger computers.

COMPUTER SOFTWARE

Software consists of the programs that enable and instruct the hardware (computer and peripheral devices such as printers) to perform desired tasks. Computer programs are categorized as either systems software or applications software. This book deals with applications software. Systems software regulates the operations of the computer and serves as an interface between applications programs and hardware. The operating system is the most important component of the system software. The operating system for microcomputers is called the Disk Operating System (DOS). The DOS allows the user to control the computer and its peripherals, manage files, and run other software.

Operating systems can be categorized as either single- or multiple-tasking systems and as either single- or multiple-user systems. With a multitasking system, more than one program can be run at the same time. The DOS distributes central processing unit (CPU) time among the different applications. With a multiuser system, more than one user can run an application. MS-DOS has become a standard for single-user, single-tasking microcomputer systems. MS-DOS for IBM compatibles is a generic version of PC-DOS for IBM personal computers. The latest versions of Unix are state-of-the-art operating systems for multiple-user environments. The Unix-based Xenix is a multiuser, multitasking system for microcomputers, including 30286-, 386-, and 486-based machines. IBM OS/2 is another multitasking operating system.

Windows, introduced by Microsoft in 1990, is a graphical environment used with MS-DOS. Windows is an extension of DOS that provides a Graphical User Interface (GUI) and multitasking. Its menus, icons (meaningful symbols), and dialog boxes replace the often cryptic MS-DOS command structure. Windows allows several application programs to run at the same time, and information can be exchanged easily between applications. With Windows, an IBM-compatible desktop running under MS-DOS provides a user environment somewhat similar to the popular Apple microcomputers. Applications software must be specifically programmed and compiled to utilize the graphical interface and other features of Windows.

IBM-compatible desktop computers operating under MS-DOS became popular in the water management community and elsewhere during the 1980s. Most of the programs cited in this book are available in executable form for MS-DOS-based desktop computers. However, the source codes can be compiled to run on other machines under other operating systems just as well. Most of the models have been run on various computer systems.

Applications programs include program development software; commercially available applications programs used in many fields (including water management); and models developed specifically for water management applications. Program development software includes assembler and higher-level languages such as FORTRAN, C, C++, and BASIC (discussed later in this chapter), which were used to code the water management models cited in Chapters 3 through 10. The myriad of commercially available applications programs includes word processors, spreadsheets, database managers, equation solvers, graphics packages, communications software, and numerous other types of applications. Several such programs are discussed in Chapter 3.

The CPU of a computer manipulates data bits according to a simplified set of sequences called machine code. People write computer programs in high-level languages which must be translated to machine code. An interpreter translates high-level language instructions one at a time, for immediate execution by the CPU. Alternatively, a compiler translates an entire high-level program to machine language. The compiled program, in machine language, is commonly associated with the extension EXE in the filename of the program. Most of the models cited in this book are available in compiled or executable ready-to-run format for IBM-compatible, MS-DOS-based desktop computer systems. Source codes in the original high-level language, such as FORTRAN, are also typically available. The source code is required if the actual program code is to be modified or if the program is to be compiled for execution on another type of computer system.

Numerous computer language translation software packages are marketed for developing application programs. Although water management models have been written in a variety of high-level languages, FORTRAN dominates. FORTRAN (FORmula TRANslator) is the oldest high-level programming language, dating back to the 1950s, and continues to be used widely in engineering and science. A majority of the programs cited in this book are coded in FORTRAN77, which is a standardized version of FORTRAN approved by the American National Standards Institute (ANSI) in 1978. Many water management models originally developed in earlier versions of FORTRAN have been updated to FORTRAN77. FORTRAN77 has the advantage of a more structured programming style. FORTRAN77 compilers are available for all categories of computers. Compilers marketed by Microsoft and Lahey are often used on microcomputers. ANSI adopted FORTRAN90 in 1994 as an updated standard version to supersede FORTRAN77.

BASIC (Beginner's All-Purpose Symbolic Instruction Code) and Pascal (named after Blaise Pascal) are examples of other programming languages that have been used in water resources planning and management. These two relatively simple languages are popular for use on microcomputers. BASIC is particularly popular as an easy-to-learn, general-purpose language for nonprofessional programmers. The disadvantages of BASIC are a tendency toward long, unstructured programs that are difficult to modify and cumbersome input of text and graphics. Pascal is a highly structured language suitable for scientific and technical applications. BASIC interpreters/compilers include Microsoft's QuickBASIC, Visual Basic, and Basic Professional. Borland's TurboPascal is a popular compiler for converting programs written in Pascal to executable machine code. Microsoft also markets a popular Pascal compiler.

C and the newer object-oriented C++ are multilevel (assembly and compiler levels) programming languages that are popular with system programmers, who are the computer experts that develop system software and application packages. C and C++ provide excellent graphics capabilities and optimize computational efficiency. These languages have been used fairly extensively in recent years to develop water management models. In some cases, complete models are coded in these languages. In other cases, C or C++ are used to develop graphical user interfaces for FORTRAN77 water management models. Borland and Microsoft both sell C and C++ compilers for various computer systems.

FORTRAN, C, and Pascal are based on a traditional, structured approach to programming in which programs are organized into algorithmic procedures and control structures. The alternative approach of object-oriented programming is experiencing increasing popularity in recent years. C++ and Smalltalk are two of the more popular object-oriented programming languages (Lorenz 1993). Instructions and information are coded and stored as objects or modules. A program is treated as a collection of objects. An object-oriented programming system provides data encapsulation, in which a well-defined interface is used to access data. Methods are the interface functions that provide access to the encapsulated data. The data and methods together make up an object. Inheritance is another fundamental concept of object-oriented programming, which provides a mechanism for constructing complex objects from simpler ones. Objects can be reused in different programs and subprograms, and programs are easier to modify. The object-oriented approach also facilitates development of graphical simulation environments in which users create simulation models by graphically linking icons that represent system objects.

A computer program is in the public domain when its development has been supported through public funds, and no distribution restrictions, copyrights, or patents apply. Public domain software can be copied and distributed freely. No one can copyright public domain software. However, in some cases, private firms have significantly modified or enhanced public domain software and copyrighted the enhancements. Computer software is proprietary if a private entity owns the copyright, trademark, or patent. Distribution of proprietary software is legally subject to restrictions established by the owner of the software rights. Most of the software cited in this book is in the public domain. Public agencies normally charge a nominal (typically $100 to $200) handling fee for distributing the software, but the user can make copies without restriction. The handling fee is typically negligible relative to the agency's cost of developing the software or its value to the user. Some agencies distribute computer programs free of charge or require the requester to furnish the necessary diskettes or other storage media. Charges for printed documentation and reports are usually based on the cost of reproduction and shipping.

In many fields of engineering, science, and business, both customized and off-the-shelf computer software is extremely expensive. In many highly competitive industries, technology transfer is often severely constrained, and each firm bears the full cost of its own software development. Some expensive software is also used in water resources planning and management. However, in general, the water management community is fortunate to have ready access to a tremendous inventory of low-cost software. Water management occurs in the public sector with free sharing of technology and information. The numerous public domain software packages cited throughout this book required large amounts of public funds to develop but are now available to all interested users, with only nominal handling charges. The popular commercially available proprietary software packages such as spreadsheets, which are used widely in water management, are inexpensive because of their extremely large market in so many areas of business, science, engineering, and other professional fields.

MODELING SYSTEMS

A water management modeling application typically involves several software packages used in combination. For example, a river basin management application might involve a modeling system which includes a watershed runoff model (Chapter 7) used to develop runoff hydrographs and pollutant loadings for input to a river and reservoir water quality model (Chapter 9) and/or a reservoir/river system operation model (Chapter 10), which in turn determines discharges and contaminant concentrations at pertinent locations in the river/reservoir system. The example modeling system could also include a stream hydraulics model (Chapter 8) to compute flow depths and velocities associated with the flows from the watershed runoff and reservoir releases determined with the other models. A geographic information system, spreadsheet program, and other data management programs (Chapter 3) are included in the modeling system to (1) develop and manage voluminous input data; (2) perform statistical and graphical analyses of simulation output; and (3) display and communicate modeling results. Thus, although Chapters 3 through 10 focus on individual modeling packages, it is important to emphasize that developing a modeling system for a particular application often involves integration of several of the different types of programs.

The software incorporated into modeling systems can be categorized as follows:

- Models that simulate the real-world system
- User interfaces
- Preprocessor programs for acquiring, preparing, checking, manipulating, managing, and analyzing model input data
- Postprocessor programs for managing, analyzing, interpreting, summarizing, displaying, and communicating modeling results

Chapters 4 through 10 address models developed specifically to simulate natural and human-made real-world water resources systems. Many of the software packages include various pre- and postprocessor utility programs along with the water resources system simulation models. Although some of the software packages cited in Chapter 3 have been routinely used to develop complete water management models, others are more typically used as preprocessor and postprocessor programs.

Model development in recent years has been characterized by an emphasis on interactive user interfaces oriented toward using advances in computer technology to make models more convenient to use. Graphical user interfaces (GUI) are popular. A GUI is any user interface that has windows, icons, menus, and pointers (Hix and Hartson 1993). A GUI is particularly important for applications involving the production or processing of graphic images such as maps, diagrams, drawings, and charts. Enhanced user interfaces have been a key consideration incorporated in the development of newer water management models and have been added recently to a number of older models.

The concept of decision support systems became popular during the 1980s in the water management community as well as in business, engineering, and other professional fields in general (Johnson 1986; Thierauf 1988). A decision support system is a user-oriented computer system that supports decision makers in addressing unstructured problems. The general concept emphasizes the following:

- Solving unstructured problems which require combining the judgment of manager-level decision makers with quantitative information
- Capabilities to answer what-if questions quickly and conveniently by making multiple runs of one or more models
- Use of enhanced user-machine interfaces
- Graphical displays

Decision support systems include a collection of software packages and hardware. For example, decision support systems are used for real-time flood control operations of reservoir systems. Making release decisions during a flood event is a highly unstructured problem because reservoir operations depend on operator judgment as well as prespecified operating rules and current and forecasted streamflow, reservoir storage level, and other available data. The decision support system includes data management software (Chapter 3); watershed runoff, stream hydraulics, and reservoir/river system operation models (Chapters 7, 8, and 10); a computer platform with various peripheral hardware devices; and an automated real-time hydrologic (streamflow and rainfall) data collection system.

The water management models cited in this book are often used as components of decision support systems. The models are applied more often in other planning, design, and resource management situations that do not exhibit all the characteristics attributed to decision support systems. For example, the watershed runoff and stream hydraulics programs used in the real-time flood control operations decision support system mentioned in the previous paragraph are also applied to the much more structured problems of delineating flood plains for the National Flood Insurance Program and sizing culverts for a new highway. Essentially all models are tools for supporting various decision-making activities. However, the relationships between decision-making processes and modeling systems vary depending on the particular water management application.

REFERENCES

HENLE, R. A., and B. W. KUVSHINOFF, *Desktop Computers,* Oxford University Press, 1992.

HIX, D., and H. R. HARTSON, *Developing User Interfaces,* John Wiley & Sons, 1993.

HOPPER, G. M., and S. L. MANDELL, *Understanding Computers,* 3rd edition, West Publishing Company, 1990.

JOHNSON, L. E., "Water Resource Management Decision Support Systems," *Journal of Water Resources Planning and Management*, American Society of Civil Engineers, Vol. 112, No. 3, July 1986.

LORENZ, M., *Object-Oriented Programming, A Practical Guide,* Prentice Hall, 1993.

Office of Technology Assessment, "Use of Models for Water Resources Management, Planning, and Policy," Congress of the United States, Washington, D.C., August 1982.

THIERAUF, R. T., *User-Oriented Decision Support Systems,* Prentice Hall, 1988.

2

MODEL DEVELOPMENT AND DISTRIBUTION ORGANIZATIONS

THE MODEL DEVELOPMENT COMMUNITY

A number of software packages that are popular in many areas of business, engineering, science, education, and other professional fields are also used extensively in water resources planning and management. Several such computer programs are cited in Chapter 3. These proprietary programs are developed and marketed for commercial profit by private enterprise. The software products can be purchased from local retail stores as well as from the software development companies.

The majority of the water management models noted in this book are public domain software packages developed under the auspices of federal agencies. In some cases, the models were developed in house by agency personnel. In other cases, the models were developed by university researchers or consulting firms working under contracts with federal agencies. Model development is often an evolutionary process with various agency, university, and consulting firm personnel making contributions at various times. Public domain software packages developed and maintained by federal agencies are used widely by other federal agencies, state and local governmental entities, private consulting firms, various industries, and universities.

State and local agencies also develop public domain generalized water resources software packages, but not nearly to the extent that the federal water agencies do. The California Department of Water Resources, Illinois State Water Survey, and Texas Water Development Board are notable examples of state agencies that have been active in model development.

Numerous university researchers are active in developing and applying water management models. Universities are oriented toward development of innovative modeling concepts and technology. Universities tend to be particularly strong in developing new models. Federal agencies typically have stronger institutional capabilities for long-term maintenance and support of models. A majority of water-related university research projects are either completely funded or cost shared by federal grants and contracts.

Numerous private firms are active in water resources modeling, including engineering consulting firms and companies specializing in software development and/or marketing. Consulting firms routinely apply models in studies conducted for their clients. Many firms also distribute computer programs developed by the federal water agencies as well as other nonfederally developed software. Most of the firms that distribute software also provide various forms of technical support for model users. Private firms have added various enhancements to federal computer programs, such as graphical user interfaces.

The International Ground Water Modeling Center (Table 2.1) maintains a directory of firms and organizations that are active in development and distribution of ground-water software. Walton (1993) and Anderson et al. (1993) also provide lists of software distributors for ground-water models. The Hydrologic Engineering Center (HEC; see Table 2.1) maintains a directory of vendors that distribute HEC computer programs. The December 1993 list of HEC model distributors includes nine universities and 63 private firms. Many of these entities also distribute software other than HEC programs and provide various modeling support services.

Donley Technology (Box 335, Garrisonville, VA 22463, 703/659-1954) publishes the *Environmental Software Report* (eight issues per year) and the *Environmental Software Directory*. The 1992/1993 *Environmental Software Directory* cites more than 900 software packages, on-line systems, and databases for hazardous substance management, regulatory compliance, risk assessment, air pollution and ground-water modeling, water/wastewater management, and mapping and geophysical information from both governmental and commercial sources.

INTERNATIONAL PERSPECTIVE

This book is written from the perspective of model development and application in the United States. Most, although not all, of the models cited were developed in the United States. The United Nations (1994) provides an overview of software and hardware available for microcomputer applications in water resources planning and development, written primarily for water professionals in developing countries. A number of water management models developed in a variety of countries throughout the world are noted. The International Ground Water Modeling Center (IGWMC) is discussed later in this chapter. The IGWMC maintains offices in the United States

TABLE 2.1   Selected Federal and Federally Supported
Model Development and Distribution Organizations

Hydrologic Engineering Center
  U.S. Army Corps of Engineers
  609 Second Street
  Davis, CA 95616
  (916)756-1104

Institute for Water Resources
  U.S. Army Corps of Engineers
  Casey Building, 7701 Telegraph Road
  Alexandria, VA 22310-3868
  (703)355-3042

Waterways Experiment Station
  U.S. Army Corps of Engineers
  3909 Halls Ferry Road
  Vicksburg, MS 39180-6199
  (800)522-6937, (601)634-2581

Water Resources Division
  U.S. Geological Survey
  409 National Center
  Reston, VA 22092
  (703)648-5215

Office of Hydrology
  National Weather Service, NOAA
  1325 East-West Highway
  Silver Spring, MD 20910
  (301)713-0006

Soil Conservation Service
  U.S. Department of Agriculture
  P.O. Box 2890
  Washington, DC 20013-2890
  (202)720-4525

Bureau of Reclamation
  U.S. Department of the Interior
  Denver Federal Center, Bldg. 67
  P.O. Box 25007
  Denver, CO 80225
  (303)236-9208

Center for Exposure Assessment Modeling
  Environmental Research Laboratory
  U.S. Environmental Protection Agency
  960 College Station Road
  Athens, GA 30613-0801
  (706)546-3549

Center for Subsurface Modeling Support
  Kerr Environmental Research Laboratory
  U.S. Environmental Protection Agency
  P.O. Box 1198
  Ada, OK 74820
  (405)436-8500

International Ground Water Modeling Center
  Institute for Ground Water Research and
  Education
  Colorado School of Mines
  Golden, CO 80401-1887
  (303)273-3103

National Technical Information Service
  U.S. Department of Commerce
  5285 Port Royal Road
  Springfield, VA 22161
  (703)487-4600

McTrans Center for Microcomputers in
  Transportation
  University of Florida, 512 Weil Hall
  Gainesville, FL 32611-2083
  (904)392-3224

(Table 2.1) and at the TNO Institute of Applied Geoscience, P.O. Box 6012, 2600 JA Delft, The Netherlands.

The Hydrologic Operational Multipurpose Subprogramme (HOMS) of the World Meteorological Organization (WMO) is a program to facilitate sharing of technology used by hydrologists throughout the world. HOMS consists of the organized transfer of hydrologic technology in the form of components consisting of manuals of (1) procedures and general guidance, (2) description of equipment, and (3) computer software. HOMS facilitates the exchange of computer models, as well as other forms of hydrologic technology, between countries. HOMS is headquartered at the following address: Hydrology and Water Resources Department, WMO Secretariat, Case Postale No. 2300, CH-1211, Geneva 2, Switzerland. HOMS National Reference Centers are located in various countries. The National Weather Service (NWS) Office of Hydrology (Table 2.1) serves as the United States National Reference Center. Information regarding available water management models may be obtained by contacting the HOMS headquarters or one of the National Reference Centers.

FEDERAL AGENCIES

Several key federal and federally supported organizations which maintain and distribute generalized water management software are listed in Table 2.1 along with their addresses and telephone numbers. Numerous water management models, including a majority of the models cited in this book, can be obtained by contacting these organizations. The remainder of this chapter provides a discussion of the federal agencies and federally supported software distribution organizations listed in Table 2.1.

U.S. ARMY CORPS OF ENGINEERS

The U.S. Army Corps of Engineers (USACE) is the nation's oldest and largest water resources development and management agency. The nationwide civil works mission of the USACE includes planning, design, construction, operation, and maintenance of facilities for navigation, flood control, water supply, hydroelectric power generation, recreation, protection and enhancement of environmental resources, and water quality management, as well as other resource management and regulatory functions. The USACE division and district offices are active in computer modeling, and several have developed generalized models that are used in other non-USACE as well as USACE offices. However, the following discussion focuses on the Hydrologic Engineering Center (HEC), Institute for Water Resources (IWR), and Waterways Experiment Station (WES), which are support organizations that provide a range of services to the field offices, including computer modeling support.

Hydrologic Engineering Center

The USACE HEC was established in 1964 to develop generalized computer programs and related technical support services for the field offices involved in the USACE water resources development program. HEC activities in hydrologic engineering and planning analysis include applied research, development of analysis methods, provision of short courses and other training, and assistance for field offices with special studies. The HEC has a professional staff of about 25 engineers and computer scientists plus a number of support personnel. Over the years, the HEC has developed numerous computer programs. Many are used widely by other agencies, consulting firms, and universities, as well as by USACE offices. Currently available major HEC software packages are listed in Table 2.2. Feldman (1981) and Davis and Bonner (1990) provide general overviews of HEC models and model development, distribution, and support activities.

The HEC-1 Flood Hydrograph Package and HEC-2 Water Surface Profiles programs, which are discussed in Chapters 7 and 8, are particularly notable examples of federally developed computer models which are used extensively by private consulting firms and universities as well as water agencies. HEC-1 and HEC-2 have been adopted by the practicing engineering community as standards for basic watershed runoff, flood routing, and backwater computations. The widespread use of HEC-1 and HEC-2 outside of the USACE has been motivated largely by the flood-plain studies required by the National Flood Insurance Program and associated local flood-plain management activities.

A HEC computer program catalog, list of model distributors (vendors), and publications catalog are available by contacting the HEC. HEC publications, including computer program documentation, can be ordered directly from the HEC using the price list and order form provided in the catalog. The more popular HEC programs are distributed on diskette through private vendors and the National Technical Information Service (NTIS). Any HEC programs not available from the NTIS or private vendors can be obtained directly from the HEC. Federal agencies can obtain any of the programs directly from the HEC.

Most of the HEC programs were originally developed for mainframe and minicomputer systems. In recent years, most of the programs have also become available in executable format for IBM-compatible microcomputers using MS-DOS. The programs are written in FORTRAN77. The HEC is also pursuing the use of workstations operating under Unix as an additional major environment for its models.

The HEC conducts several one- and two-week short courses each year based on the generalized simulation modeling packages. The courses are primarily for USACE personnel, with non-USACE personnel admitted on a space-available basis. A number of universities and consulting firms also offer short courses on HEC models, particularly HEC-1 and HEC-2. Numerous publications are available from the HEC, including computer program users manuals, training documents, and reports and papers on specific applications of the models.

Institute for Water Resources

The IWR is part of the USACE Water Resources Support Center located at Fort Belvoir, Virginia. The IWR mission is to analyze and anticipate changing water resources management conditions and to develop planning methodologies to address

TABLE 2.2 USACE Hydrologic Engineering Center Models

AGDAM	Agricultural Flood Damage Analysis
COED	Corps Editor
DSSMATH	Mathematical Utilities for DSS Data
HEATX	Heat Exchange Program
HEC-1	Flood Hydrograph Package
HEC-1F	Modified HEC-1 for Real-Time Water Control Systems
HEC-1FH	Interior Flood Hydrograph Package
HEC-2	Water Surface Profiles
HEC-4	Monthly Streamflow Simulation
HEC-5	Simulation of Flood Control and Conservation Systems
HEC-5Q	Simulation of Flood Control and Conservation Systems with Water Quality Analysis
HEC-6	Scour and Disposition in Rivers and Reservoirs
HECDSS	Data Storage System
HEC-FDA	Flood Damages Analysis Package
HEC-FFA	Flood Frequency Analysis
HEC-LIB	HEC Subroutine Library
HEC-PRM	Prescriptive Reservoir Model
HGP	Hydraulics Graphics Package
HMR52	Probable Maximum Storm (Eastern United States)
HYCOST	Small-Scale Hydroelectric Power Costs Estimates
HYDPAR	Hydrologic Parameters
HYDUR	Hydropower Analysis Using Streamflow Duration Procedures
MLRP	Multiple Linear Regression Program
NWSDSS	Load NWS Data Tapes in DSS
PAS	Preliminary Analysis System for Water Surface Profile Computations
REGFRQ	Regional Frequency Computation
RESTMP	Reservoir Temperature Stratification
RMA-2	Finite Element Hydrodynamics
STATS	Statistical Analysis of Time Series Data
STORM	Storage, Treatment, Overflow, Runoff Model
THERMS	Thermal Simulation of Lakes
UHCOMP	Interactive Unit Hydrograph and Hydrograph Computation
UNET	One-Dimensional Unsteady Flow Through a Full Network of Open Channels
WQRRS	Water Quality for River-Reservoir Systems
WQSTAT	Water Quality Statistics
WATDSS	Load WATSTORE Data in DSS

Water Control Programs (includes a number of programs for real-time water control) economic, social, institutional, and environmental needs in water resources planning and policy. IWR develops tools and strategies needed to plan and execute USACE water resources programs. IWR activities and products also serve the overall water management community. This publication is an IWR product. The IWR-MAIN water use forecasting model discussed in Chapter 4 is another IWR product.

IWR was created in 1969. Its permanent staff of about 40 employees is supplemented by contract research and temporary assignments of professionals from other Corps offices, other agencies, universities, and consulting firms. The IWR organization structure includes four divisions: Technical Analysis and Research, Policy and Special Studies, Navigation, and Program Analysis.

Waterways Experiment Station

The U.S. Army Engineer WES is a Corps of Engineers research complex consisting of six laboratories: Coastal Engineering Research Center, Environmental Laboratory, Geotechnical Laboratory, Hydraulics Laboratory, Structures Laboratory, and Information Technology Laboratory. WES was established in 1929, after the disastrous Mississippi River Flood of 1927, and is the largest research facility within the Corps of Engineers. The facility is located on 685 acres in Vicksburg, Mississippi. A professional staff of over 700 engineers and scientists perform research, test materials and equipment, and provide consulting services to the USACE field offices, the U.S. Army and Air Force, and other federal agencies.

Numerous computer models have been developed at WES over the years for studies sponsored by the USACE field offices and other federal agencies. The *WES Computer Program Library Catalog* lists several hundred programs developed by the six laboratories. Modeling applications at WES are typically for specific projects. However, WES has many generalized models which are available to the public. Information regarding requests for models and related documentation can be obtained by contacting the WES Engineering Computer Programs Library at the telephone number and address provided in Table 2.1. Information regarding specific models can also be obtained by contacting technical personnel in the pertinent laboratories.

U.S. GEOLOGICAL SURVEY

The mission of the Water Resources Division (WRD) of the U.S. Geological Survey (USGS) is to provide the hydrologic information and understanding needed to support management and use of the nation's water resources. The USGS WRD conducts three major types of interrelated activities: (1) data collection and dissemination, (2) problem-oriented water resources appraisals and interpretive studies, and

(3) research. The USGS WRD organizational structure includes 48 district offices generally located in state capitals and four regional offices.

The USGS WRD is active in developing water management models, including both ground-water and surface-water models, involving both water quality and quantity. A number of generalized operational models are available from the USGS WRD, through either its headquarters in the USGS National Center in Reston or the regional and district offices.

In addition to actually developing models, most of the activities of the USGS WRD are pertinent to modeling from various perspectives. For example, USGS data collection programs provide water quality and quantity data for streamflow, reservoir storage, and ground water which are essential to modeling studies. All data collected are stored in the computer-based WATSTORE (WATer data STOrage and REtrieval) system as well as being published in printed reports. Data are available upon request from WATSTORE in machine-readable format or as computer-printed tables or graphs, statistical analyses, and digital plots. As another example, the Water Resources Scientific Information Center abstracts water-resource publications from throughout the world and makes this information available to the water management community and the public through publications and computerized bibliographic information services. A comprehensive collection of abstracts for water-related publications can be searched conveniently from a CD-ROM (compact disc read-only memory).

NATIONAL WEATHER SERVICE

The National Weather Service (NWS) is responsible for the hydrologic services programs of the National Oceanic and Atmospheric Administration (NOAA) of the Department of Commerce. Development and application of hydrologic models within the NWS is associated largely with the activities of the River Forecast Centers. The NWS River Forecast Centers are responsible for issuing weather and river forecasts and warnings. The NWS hydrologic forecasts are used by other federal, state, and local entities responsible for reservoir operations, flood control, water supply management, hydroelectric power generation, and other water management functions. The NWS river forecasting activities rely heavily on computer models. The NWS Hydrologic Research Laboratory (of the Office of Hydrology located in Silver Spring, Maryland) has developed a number of models that are used widely outside the NWS as well as by the several NWS River Forecast Centers. As previously noted, the NWS Office of Hydrology also serves as the United States National Reference Center for the HOMS of the World Meteorological Organization.

SOIL CONSERVATION SERVICE

The Soil Conservation Service (SCS) of the U.S. Department of Agriculture (USDA) conducts national programs dealing with soil and water conservation and small watershed flood protection. The SCS works closely with farmers, ranchers, land-owners, and state and local organizations in carrying out its programs. The SCS has a background in computer modeling dating back to the 1950s. SCS generalized models are used widely both within and outside the agency. Technical Releases TR-20 and TR-55, which deal with watershed hydrology, are used widely by nonfederal as well as federal entities. The SCS distributes its models to the public through the NTIS. Other federal agencies can obtain models directly from the SCS. Technical assistance is provided by the office of the SCS State Conservation Engineer in each state.

BUREAU OF RECLAMATION

The U.S. Bureau of Reclamation is involved in development and management of water resources in the 17 western states for multiple purposes, including irrigation, hydroelectric power, municipal and industrial water supply, pollution abatement, propagation of fish and wildlife, recreation, drainage, erosion control, and flood control. The Bureau recently published an inventory of hydrologic models (U.S. Bureau of Reclamation 1991), which includes the categories of (1) water require-ments models, (2) precipitation-runoff models, and (3) project and river system operations models. The first phase of preparing this inventory focused on models maintained and supported by the Bureau of Reclamation which could be of interest to offices within and outside the agency. A limited number of models available from other organizations are also included in the inventory.

ENVIRONMENTAL PROTECTION AGENCY

The responsibilities of the U.S. Environmental Protection Agency (EPA) include establishing and enforcing environmental standards, conducting research on the impacts of pollution and ways to control it, and advising the President and Congress on matters of environmental policy. The EPA Office of Research and Development operates 14 national research laboratories. The Center for Exposure Assessment Modeling (CEAM) at the Environmental Research Laboratory in Athens, Georgia, and the Center for Subsurface Modeling Support (CSMoS) at the Robert S. Kerr Environmental Research Laboratory in Ada, Oklahoma, are particularly pertinent to our topic of water management models.

Center for Exposure Assessment Modeling

The CEAM was established in 1987 to meet the scientific and technical exposure assessment needs of the EPA program and regional offices and of state environmental agencies. To support environmental risk-based decisions concerning protection of air, water, and soil, CEAM provides proven predictive exposure assessment techniques for aquatic, atmospheric, terrestrial, and multimedia pathways for organic chemicals and metals. A wide range of analysis techniques is provided, from simple calculator techniques suitable for screening analysis through computerized steady-state models to sophisticated state-of-the-art continuous simulation models. The inventory of CEAM computer programs is listed in Table 2.3. These models include simulation of urban and rural nonpoint sources, conventional and toxic pollution of streams, lakes, and estuaries; tidal hydrodynamics; geochemical equilibrium; and aquatic food chain bioaccumulation. CEAM has a staff of over 30 professionals from a wide variety of disciplines and backgrounds. The CEAM staff is responsible for the analysis, design, development, distribution, and support of CEAM models.

Information regarding CEAM models, documentation, and distribution procedures can be obtained by telephoning or writing the Center. CEAM maintains versions of its software for both IBM-compatible MS-DOS-based microcomputers and DEC VAX/VMS minicomputer systems. Most programs are coded in FORTRAN77. Computer programs can be requested either directly from the CEAM by mail or through the CEAM Electronic Bulletin Board System (BBS).

TABLE 2.3   EPA CEAM Models

ANNIE-IDE	Interactive Development Environment
CLC	Coordinated List of Chemicals Data Base
CORMIX	Cornell Mixing Zone Expert System
DBAPE	Data Base Analyzer and Parameter Estimator
EXAMSII	Exposure Analysis Modeling System II
FGETS	Food and Gill Exchange of Toxic Substances
GCSOLAR	Green Cross Solar Program
HSPF	Hydrologic Simulation Program—FORTRAN
LC50	LC50 Values Estimation Program
MINTEQA2	Geochemical Equilibrium Speciation Model
MULTIMED	Multimedia Exposure Assessment Model
PRZM2	Pesticide Root Zone Model
QUAL2E	Enhanced Stream Water Quality Model
SWMM	Storm Water Management Model
WASP5	Water Quality Analysis Simulation Program
DYNHYD5	Hydrodynamic Model

The CEAM began operation of its Electronic BBS in 1988. The CEAM BBS serves four main purposes:

- The downloading of CEAM-supported software products
- The uploading of user input data sets for staff review and trouble-shooting assistance
- The dissemination of current information concerning CEAM activities and events, including announcements for CEAM workshops and training sessions, model version and update information, helpful hints for model use, and model documentation
- The rapid exchange of information between users and CEAM personnel concerning model use, problems, and enhancements

To communicate with the CEAM BBS, the user must have a microcomputer, modem, and telecommunication software package. This equipment is relatively inexpensive and readily available from local computer stores. No previous experience with the use of electronic bulletin boards is required as long as the user is familiar with his or her telecommunication software and modem.

Center for Subsurface Modeling Support

The Robert S. Kerr Environmental Research Laboratory serves as the EPA's Center for Ground-Water Research, with a focus on studies of (1) the transport and fate of contaminants in the subsurface, (2) the development of methodologies for protection and restoration of ground-water quality, and (3) the evaluation of the applicability and limitations of using natural soil and subsurface processes for the treatment of hazardous waste. The CSMoS is part of the Technology Support Center of the Kerr Laboratory. The CSMoS distributes and services all models and databases developed by the Kerr Laboratory and provides general support on model application to ground-water and vadose zone problems. Software and support services are provided to public agencies and private companies throughout the nation. The models listed in Table 2.4 are available upon request from the CSMoS.

Exposure Models Library and Integrated Model Evaluation System

The Exposure Assessment Group, Office of Health and Environmental Assessment, in the Environmental Protection Agency's Office of Research and Development (Washington, D.C. 20460, 202/260-8922) recently developed the Exposure Models Library and Integrated Model Evaluation System (EML/IMES). The EML/IMES

TABLE 2.4   EPA CSMoS Models

BIOPLUME II	Two-Dimensional Contaminant Transport Under the Influence of Oxygen Limited Biodegradation in Ground Water
CHEMFLO	One-Dimensional Water and Chemical Movement in Unsaturated Soils
GEOPACK	Geostatistics for Waste Management
MOFAT	Two-Dimensional Finite Element Program for Multiphase Flow and Multicomponent Transport
MT3D	Modular Three-Dimensional Transport Model
OASIS	Parameter Estimation System for Aquifer Restoration Models
PESTAN	Pesticide Analytical Model
RETC	Retention Curve Computer Code
RITZ	Regulatory and Investigative Treatment Zone Model
STF	Soil Transport and Fate Database 2.0 and Model Management System
VLEACH	Vadose Zone Leaching Model
WHPA	Modular Semi-Analytical Model for the Delineation of Wellhead Protection Areas

(EPA/600/C-92/002) is distributed to users on CD-ROM. The CD-ROM disc contains (1) the Integrated Model Evaluation System (IMES), with information on selecting an appropriate model, literature citations on validation of models in actual applications, and a demonstration of a model uncertainty protocol; (2) the Exposure Models Library (EML), which includes the model source codes, sample input and output data, and, in many cases, model documentation; and (3) a variety of other information, including catalogs, directories, and indexes of EPA programs and activities and environmental information resources. The September 1993 version of the Exposure Models Library contains over 90 models (including those listed in Tables 2.3 and 2.4), developed by the EPA and others, for determining fate and transport in various environmental media (air, ground water, surface water, soil).

INTERNATIONAL GROUND WATER MODELING CENTER

The International Ground Water Modeling Center (IGWMC) was established in 1978 at the Holcomb Research Institute of Butler University in Indianapolis, Indiana. In 1991, the IGWMC was relocated to the Colorado School of Mines in Golden, Colorado. The European office of IGWMC began operation in 1984 and is located at the TNO Institute of Applied Geoscience, Delft, The Netherlands. The activities of the United States office of the IGWMC have been supported largely by the EPA, through its Robert S. Kerr Environmental Research Laboratory. The IGWMC operates a clearinghouse for ground-water modeling software. Models available through the IGWMC are tabulated in Chapter 6. The IGWMC provides a catalog of available models upon request.

IGWMC activities include information dissemination, software distribution, research, training and education, and communications. The modeling software distribution and support activities include model evaluation, code acquisition and implementation, code testing, preparation of documentation, and development of code support activities. Models undergo a quality-assured implementation procedure before being released by the Center. In addition to acquisition and distribution of ground-water model codes, the IGWMC develops and distributes modeling-based decision support systems, including pre- and postprocessors for data preparation, simulation software, and textual and graphical displays of simulation results. IGWMC also maintains a research collection of program documentation reports and modeling-related literature, and it conducts an ongoing program of applied research and development. Topics covered in the research program include quality assurance in modeling, model screening and testing, evaluation of model use and model needs, and model review studies. IGWMC offers annual short courses, workshops, and seminars and provides assistance to government agencies and private groups in organizing and conducting training programs. The Center distributes a quarterly *Ground Water Modeling Newsletter.*

TABLE 2.5 Selected Water-Related Microcomputer Programs
Distributed by National Technical Information Service

USDA Soil Conservation Service Models	
TR-20	Computer Program for Project Hydrology
TR-20	Interactive Input Computer Program
TR-55	Urban Hydrology for Small Watersheds
USACE Hydrologic Engineering Center Models	
COED	Corps Editor
EAD	Expected Annual Flood Damage Computation
HEC-1	Flood Hydrograph Package
HEC-2	Water Surface Profiles
HEC-5	Simulation of Flood Control and Conservation Systems
HEC-6	Scour and Deposition in Rivers and Reservoirs
HECWRC	Flow Frequency Analysis
HMR52	Probable Maximum Storm (Eastern United States)
HYCOST	Small-Scale Hydroelectric Power Cost Estimates
MLRP	Multiple Linear Regression Program
PAS	Profile Accuracy System
STATS	Statistical Analysis of Time Series Data
WQRRS	Water Quality for River/Reservoir Systems

TABLE 2.6 Hydraulics Microcomputer Programs Distributed by McTrans

ASHDRAIN	Design of Inlets and Drainage Networks
CodeH2	Expert System for HEC-2
Culvert Analysis (HY-8)	
DABRO	Drainage Basin Runoff Model
DAMP	Drainage Analysis and Modeling Programs
EASy	Engineering Analysis System
FESWMS	Finite Element Surface Water Modeling System
FLOWMASTER	
HC2ENTRY	HEC-2 Input
HEC-1	Flood Hydrograph Package
HEC-2	Water Surface Profiles
HEC-12	FHWA Hydraulic Circular 12 (Pavement Drainage)
HYDGEN	Watershed Hydrographs
HYDRAIN	Highway Drainage
HydroCAD	Computer Aided Design for Hydrology and Hydraulics of Stormwater Runoff
HY-EDIT	HEC-1 and HEC-2 Edit Program
Hydrogen Sulfide	
HY-TB	Hydraulic Toolbox
LCA	Least Cost Analysis
Mac Culvert	
MacStorm Sewer	
MNDOT.HYD	Box Culvert Analysis
Preliminary Analysis System	
Scour at Bridges	
Storm Sewer Analysis and Design	
Storm Sewer Analysis and Design Utilizing Hydrographs	
SWATER	Optimal Sewer Design Package
SWITCH	HEC-2 and WSPRO Utility Program
TR20-88	Computer Program for Project Hydrology
Urban Stormwater Management Planning and Design	
VaMP	Virginia Groundwater Mounding Program
WSPRO	Water Surface Profile
WSPRO GRAPH	

NATIONAL TECHNICAL INFORMATION SERVICE

The NTIS of the U.S. Department of Commerce distributes research and development results for projects sponsored by the U.S. and foreign governments, in the format of summary announcements and technical reports. NTIS also manages the Federal Computer Products Center, which distributes software and documentation to the public for federal agencies. Table 2.5 lists Hydrologic Engineering Center and Soil Conservation Service water-related microcomputer programs distributed through the NTIS. The NTIS also distributes an assortment of other federal agency computer programs which are either directly water related or are general-purpose software which can be used in water management as well as in other fields. NTIS only distributes the models. No technical support is provided.

McTRANS CENTER FOR MICROCOMPUTERS IN TRANSPORTATION

The Center for Microcomputers in Transportation (McTrans) is a software distributor and user support center established by the Federal Highway Administration and supported by the Federal Transit Administration. The McTrans Center provides support to microcomputer users by technical assistance for the software it distributes. Although the McTrans Center focuses on highway transportation and mass transit software, several of the hydraulics models it distributes are pertinent to water management. Hydraulics software included in the June 1993 *McTrans Catalog* is listed in Table 2.6. Other general-interest software, such as spreadsheets and database managers, is also pertinent to water management. The computer programs distributed by McTrans were developed by a variety of entities, including federal and state agencies, universities, and private firms.

REFERENCES

ANDERSON, M. P., D. S. WARD, E. G. LAPPALA, and T. P. PRICKETT, "Computer Models for Subsurface Flow," in *Handbook of Hydrology,* edited by D. R. Maidment, McGraw-Hill, 1993.

DAVIS, D. W., and V. R. BONNER, "Twenty-Five Years of Developing, Distributing, and Supporting Hydrologic Engineering Center Computer Programs," *Transferring Models to Users Symposium Proceedings,* American Water Resources Association, Bethesda, Maryland, 1990.

FELDMAN, A. D., "HEC Models for Water Resources System Simulation," in *Advances in Hydroscience,* Vol. 12, edited by Ven T. Chow, Academic Press, 1981.

UNITED NATIONS, "The Use of Microcomputers in Water Resources Planning and Development," Water Resources Branch, Department of Support and Management Services, New York, January 1994.

U.S. BUREAU OF RECLAMATION, "Inventory of Hydrologic Models," Global Climate Change Response Program, Denver, Colorado, August 1991.

WALTON, W. C., *Groundwater Modeling Utilities,* Lewis Publishers, 1993.

3

GENERAL-PURPOSE SOFTWARE

INTRODUCTION

The remainder of this software guide covers various categories of computer programs. This chapter is different from subsequent chapters. The generalized models addressed in Chapters 4 through 10 are designed specifically for water management applications. Each of Chapters 4 through 10 covers a particular category of water management models. Chapter 3 deals with general-purpose software which is pertinent to several or all of the types of water management modeling applications covered in the subsequent chapters and may not necessarily be limited to water management applications. Many of the general-purpose computer programs are intended for a broad range of applications in business, engineering, science, and other professional fields. Most of the models cited in Chapters 4 through 10 are in the public domain. Although several public domain programs developed by federal agencies are covered in Chapter 3, much of the software cited is proprietary and is commercially available through software retailers or directly from the developer.

The software addressed in this chapter includes spreadsheets, object-oriented simulation modeling environments, expert systems, mathematical and statistical programs, data management systems, geographic information systems, computer-aided drafting and design, and graphics. A water management modeling application may involve use of several of these software packages in combination. The general-purpose computer programs of Chapter 3 may be used either in lieu of or in combination with the models of Chapters 4 through 10. A complete model, involving any of the water management application areas addressed in Chapters 4 through 10, may be developed using a spreadsheet, simulation modeling environment, expert system software, or mathematical programming. Spreadsheet, data management, geographic information system, graphics, and statistical programs often serve as preprocessors to manipulate input data for the water management models of Chapters 4 through 10 and/or as postprocessors to analyze, interpret, summarize, display, and communicate simulation results.

Developing and marketing computer software is a multibillion-dollar per year industry. A review of the various software directories and databases demonstrates the tremendous diversity and amount of software available. For example, *Data Sources, the Complete Computer Product Book* is published twice a year by the Ziff-Davis Publishing Company (One Park Avenue, New York, NY 10016, 212/503-3500). The 1994 edition includes citations for 75,000 hardware, software, and communications products, which are produced by 14,000 companies. The 1993 edition of *The Software Encyclopedia,* which is published annually by R. R. Bowker (121 Chanlon Road, New Providence, NJ 07974), contains entries for over 16,000 microcomputer software packages from 4,000 publishers. These and other similar directories are oriented toward proprietary software marketed by private enterprise for profit. Most of the public domain water management models are not included in the directories.

A tremendously large inventory of off-the-shelf software is available through local retail stores and mail-order vendors and directly from the software development companies. Software products are dynamic, with new versions being released periodically. The software business is extremely competitive, with similar products available from different companies, particularly in the more popular applications areas such as word processors and spreadsheets. Some applications software, such as word processing programs, is routinely used by water management agencies and firms but is only indirectly related to modeling. A significant amount of commercially available software can be directly applied in modeling water management problems. Water management professionals will continue to discover new uses for old software as well as for new products being marketed.

The inventory of general-purpose software available to the water management community is much too large and dynamic to cover in this chapter in any degree of depth or comprehensiveness. This chapter represents an attempt to outline several types of software packages which are particularly pertinent to water management modeling applications. The intent is not to evaluate or compare software available from different companies. Citation of a particular computer program does not imply endorsement of that program over another competing product sold by another company. Even more importantly, neglecting to mention a particular software product certainly does not reflect negatively on the product. Rather the intent is to highlight general categories of software which may be useful in water management modeling applications and to note the availability of several representative programs in each category.

SPREADSHEET/GRAPHICS/DATABASE SOFTWARE

VisiCalc, developed in 1978, was the first electronic spreadsheet. The ever-popular Lotus 1-2-3 was first introduced in 1982. Other spreadsheet/graphics/database programs entered the market at about the same time or soon thereafter. The competing packages evolved through various versions and still continue to be improved and expanded. Their popularity grew rapidly. Spreadsheet packages are now used routinely in millions of businesses, professional offices, universities, and homes throughout the nation and world.

Lotus 1-2-3 is so named because it provides three types of capabilities: spreadsheet computations, graphics, and database management. Lotus 1-2-3 and similar packages are commonly called spreadsheets even though all three functions are performed. The electronic spreadsheet has a matrix format of rows and columns, much like a traditional financial spreadsheet developed on paper with a pencil. A cell is the intersection of a row and column. Values stored in specified blocks of cells can be plotted in various formats. Database management is based on storing data in blocks of cells. Computations are performed using data stored in individual cells, rows, columns, or otherwise defined blocks of cells. Computational results are also stored in blocks of cells. The user programs mathematical equations and arithmetic computations for specific applications using operations and functions provided by the software. Statistical analyses can be performed. Spreadsheets also include special routines for computational tasks such as multiple linear regression analysis, solving systems of linear equations, and linear programming.

Numerous spreadsheet/graphics/database programs are on the market. Lotus 1-2-3, Quattro Pro, and Excel are among the more popular packages. Information regarding these three spreadsheet programs can be obtained by contacting the software companies at the addresses and telephone numbers provided in Table 3.1. The programs are available from most local software stores at retail prices of about $500 for microcomputer versions. Lotus 1-2-3, Quattro Pro, and Excel have similar capabilities. Worksheets are interchangeable between the programs. A worksheet developed with one program can be used with another program. The menu structures are also similar. A number of other programs on the market also replicate the basic features of these popular programs. Versions of Lotus 1-2-3, Quattro Pro, and Excel are available for IBM-compatible personal computers running MS-DOS, both with and without Windows, and for Apple Macintosh desktop computers. Versions of

TABLE 3.1   Spreadsheet Programs

Lotus 1-2-3	Lotus Development Corporation 55 Cambridge Parkway Cambridge, MA 02142 (800)343-5414, (617)577-8500
Quattro Pro	Borland International, Inc. P.O. Box 660001, 1800 Green Hills Road Scotts Valley, CA 95067-0001 (800)331-0877, (408)438-8400
Excel	Microsoft Corporation One Microsoft Way Redmond, WA 98052-6399 (800)426-9400, (206)882-8080

Lotus 1-2-3, in particular, are also available for several minicomputer and mainframe computer systems.

Spreadsheet programs have been used in conjunction with each of the major categories of water management applications covered in Chapters 4 through 10, including evaluation of water demands and supplies, water distribution system analysis, ground water, watershed runoff, stream hydraulics, river and reservoir water quality, and reservoir/river system operations. Spreadsheet packages are commonly used as preprocessors for preparing and organizing input data for the models cited in Chapters 4 through 10 and as postprocessors for summarizing, plotting, and performing statistical and other analyses of simulation results. For relatively simple applications, spreadsheets are used to develop complete models, without using the models of Chapters 4 through 10.

Spreadsheet programs have the advantage of applying the same familiar software to many different applications. A particular modeling problem can be addressed using software which is already being used in the office for other purposes as well.

Water management professionals have recognized the potential of electronic spreadsheets since soon after they were first marketed. Most applications are routine and thus never reported in the published literature. A number of spreadsheet applications in water resources have been published in the literature. Several examples are cited here to illustrate the broad range of applications addressed.

Olsthoorn (1985) pointed out the usefulness of the new so-called spreadsheet programs in ground-water modeling. Highland (1987) demonstrated the capabilities of spreadsheets in solving ground-water problems with two examples: (1) simulation of unconfined flow with infiltration, and (2) two-fluid flow with a dense fluid and a light fluid as often found in organic contamination of an aquifer. Hancock and Heaney (1987) also noted the powerful computational capabilities of spreadsheets as illustrated by several surface water applications, including database management, simple computer mapping, spatial data analysis, hydrologic budget, trend analysis, precipitation-runoff simulation, flood routing, and surface–ground water simulation.

The city of Seattle has developed a microcomputer database management system to support the design process for wastewater treatment plants, which includes Lotus 1-2-3 for data manipulation, dBASE software for storing data, AutoCAD software for preparing drawings, and Microsoft Word for word processing (Samstag et al. 1989). The city of Kalamazoo, Michigan used Lotus 1-2-3 to develop financial planning and rate models for its water supply and wastewater collection and treatment system (Barnett-Moore et al. 1990). Miles and Heaney (1988) present a stormwater drainage design method developed using Lotus 1-2-3. Walker et al. (1989) applied a spreadsheet watershed model in support of nonpoint pollution management. The Tennessee Valley Authority incorporated a spreadsheet as the data logging component of a multitasking microcomputer system used to monitor operations of a complex system of 30 hydroelectric power plants (Giles et al. 1990). Salgeonker (1989) used a spreadsheet program to evaluate water savings and the relative cost-effectiveness of various water conservation measures. Macy and Maddeus (1989) also used a spreadsheet program to perform benefit-cost evaluations of water conservation programs. Southwood et al. (1989) used Lotus 1-2-3 to model the fate of chemicals in an aquatic environment. Feistul (1990) developed a spreadsheet model to simulate water quality in a chain of three lakes subject to pollution from runoff from urbanizing watersheds. Meyer (1991) described a spreadsheet model developed to size the air flow rate supplied by a destratification system used to improve water quality in a reservoir by reducing thermal stratification. Bradley et al. (1991) developed a spreadsheet model to help determine instream flow needs to protect salmon.

OBJECT-ORIENTED SYSTEM SIMULATION SOFTWARE

A number of commercially available computer software packages are designed to provide a modeling environment or set of tools for model building. Users construct a model for their particular application within the general framework of modeling capabilities provided by the software package. The user does some programming of computational tasks, using the operations and functions provided, but the programming is much simpler than coding a model using an actual programming language such as FORTRAN, BASIC, or C. The spreadsheet programs discussed earlier provide such a modeling environment. The software packages described next are designed to simulate dynamic (time varying or otherwise changing) systems characterized by interrelated components.

STELLA II and EXTEND (Version 2) are object-oriented simulation modeling environments available from High Performance Systems and Imagine That, respectively, at the addresses and telephone numbers cited in Table 3.2. High Performance Systems also markets a version of STELLA under the name ITHINK. EXTEND and STELLA (or ITHINK) run on Apple Macintosh microcomputers. An MS-DOS Windows version of STELLA is also being developed. STELLA and EXTEND each retail for $695. An enhanced version of EXTEND, called the EXTEND+Manufacturing Package and priced at $990, provides additional capabilities for applying operations research-related mathematical methods. Printed manuals documenting STELLA and EXTEND and providing examples of applications are distributed along with the programs.

STELLA and EXTEND are general-purpose, object-oriented, system simulation modeling packages pertinent to a broad range of applications in education, business, science, engineering, and other professional fields. The user builds a simulation model for a particular application, using the modeling tools provided, and also designs the tabular and/or graphical presentation of simulation results. The models are graphically oriented and rely on standard Macintosh operations involving extensive use of a mouse.

STELLA II is an acronym for Systems Thinking, Experiential Learning Laboratory, with Animation, Version 2. A model of a system is developed using STELLA by combining four types of icons or objects: stocks, flows, converters, and connectors. Stocks accumulate flows and are used as state variables to reflect time-varying (dynamic) characteristics of the system. Numerical integration methods are used to solve the mass (volume) balance at each stock. The value or amount associated with a stock can change in each time period in response to flows into and out of the stock. For example, if a reservoir system is being modeled, stocks can represent reservoir storage, which is a time-varying function of STELLA flow objects representing stream inflows, water supply diversions, reservoir releases, and evaporation. Converters are used to store mathematical expressions and data. Con-

**TABLE 3.2   Object-Oriented Simulation
Software**

STELLA II	High Performance Systems, Inc.
	45 Lyme Road, Suite 300
	Hanover, NH 03755
	(800)332-1202, (603)643-9636
EXTEND	Imagine That, Inc.
	6830 Via Del Oro, Suite 230
	San Jose, CA 95119
	(408)365-0305

nectors provide a mechanism to indicate the linkages between stocks, flows, and converters. A system representation may consist of any number of stocks, flows, converters, and connectors. STELLA provides a number of built-in functions, which are used in developing the logic and mathematics for the particular application.

Likewise, the EXTEND graphical iconic modeling system provides libraries of objects or blocks for use in building a model for a particular application. The blocks include various computational, logical decision-making, data manipulation, and display capabilities. The user selects and connects blocks as needed to develop or modify a model. In addition to the libraries of preprogrammed blocks, EXTEND provides a built-in C-like language editor and compiler to allow the user to define customized blocks.

STELLA and EXTEND are general-purpose modeling environments pertinent to a broad range of applications in education, science, engineering, and other fields. Bogen (1989) reviewed and compared the two programs from the perspective of scientific applications. The U.S. Army Corps of Engineers (1991) suggested that STELLA and EXTEND should be adopted for application in water resources management and duplicated the watershed and reservoir simulation modeling capabilities provided by the Streamflow Synthesis and Reservoir Regulation (SSARR) model (which is described in Chapter 7), using STELLA to demonstrate its utility. Karpack and Palmer (1992) used STELLA to analyze the water supply systems of Seattle and Tacoma, Washington. STELLA was applied, in conjunction with the National Study of Water Management During Drought, in drought preparedness studies for several locations, including the Kanawha River Basin in West Virginia, James River Basin in Virginia, Cedar and Green River Basins in Washington, Marais des Cygnes-Osage River Basin in Kansas and Missouri, and the Boston, Massachusetts area (Keyes and Palmer 1993; Low and Erickson 1993; Lynch 1993; Nvule 1993; Stiles and Punnett 1993; Werick 1993). STELLA has been applied to a broad range of other problems in various disciplines, including water chemistry, biology, and ecology (Baker et al. 1991; Costanza et al. 1990; Cronk et al. 1990; Shatkin and Brown 1991).

COMPARISON OF ALTERNATIVE MODEL BUILDING APPROACHES

Having completed our review of both spreadsheet programs and object-oriented simulation software, these modeling environments are now compared with each other and with other alternative approaches. A model for a particular water management application can be constructed using either of the following sets of software tools:

- Programming languages
  - Traditional languages such as FORTRAN, BASIC, or C
  - Object-oriented languages such as C++
- Modeling environments
  - Spreadsheets such as Lotus 1-2-3, Quattro Pro, and Excel
  - Object-oriented simulation packages such as STELLA and EXTEND
- Generalized operational water management models such as those noted in Chapters 4 through 10.

Use of each set of software tools represents a different model building approach. In developing a model for a particular study, a key question is which modeling approach (or set of software tools) should be adopted. Modeling applications related to any of the general categories of water management models represented by Chapters 4 through 10 can be addressed using either of the aforementioned alternative software environments. In some situations, one alternative model building approach may be clearly advantageous over the others. In other modeling situations, the relative merits and trade-offs between the alternative sets of model building tools will be more balanced. The background and personal preferences of the model builders are typically a major consideration in selecting an approach. Several factors to be considered in comparing the relative advantages and disadvantages of the alternative approaches are discussed in the following subsections.

Key attributes of a model building environment include

- Flexibility in realistically representing the real-world system and concerns
- Capabilities which can be incorporated into the model for analyzing, displaying, and communicating simulation results
- Expertise, time, and effort required to build a model and then to apply and maintain the model

The following observations regarding the alternative approaches are presented within the context of the aforementioned attributes.

Modeling Environments versus Programming Languages

Given unlimited time, funds, and computer programming expertise, developing a program from scratch using FORTRAN, C, C++, or another similar language will provide the greatest flexibility to develop a model to fit the particular needs of the water management application. This approach also requires the greatest resources.

Thus, time and personnel resources required for detailed model construction may reduce the resources available for other more crucial aspects of the modeling study.

Lotus 1-2-3, Quattro Pro, Excel, STELLA, and EXTEND were developed by systems programmers (the real computer programming experts) working for companies able and willing to invest the resources necessary to develop polished products. These programs reflect great attention to enhanced user interfaces and graphics capabilities. They provide programming capabilities for developing computational algorithms, which are simpler than the actual programming languages such as FORTRAN, BASIC, C, and C++. The EXTEND language is somewhat similar to C. The commercial packages are very user oriented. However, in adopting these general-purpose modeling environments, water management professionals sacrifice some of the flexibility inherent in writing their own programs.

Spreadsheet Programs versus Simulation Programs

STELLA and EXTEND provide a significantly different modeling format than the spreadsheet programs (Lotus 1-2-3, Quattro Pro, and Excel). The STELLA and EXTEND system simulation environments emphasize use of icons and graphical diagrams to capture the interrelationships of system components. The distinguishing characteristics of these programs, as compared to spreadsheets, are their capabilities for graphical object-oriented representation of the interrelationships of the components of real-world systems. Convenient capabilities are also provided for repeating computations by time steps for dynamic systems. Spreadsheet programs are distinguished by being oriented to a format which facilitates organization of work in tables of user-inputted and model-computed numbers, along with associated text and labels. Spreadsheet computations are based on mathematically manipulating data stored in matrixes of rows and columns. Spreadsheets also provide graphics for plotting the data stored in blocks of rows and columns.

Both types of software have important roles in water management modeling. In many situations, models can be constructed using either approach, with the choice guided largely by the personal preferences of the users. In other cases, one approach may be more clearly advantageous than the other.

Modeling Environments versus Generalized Water Management Models

Generalized water management models such as those cited in Chapters 4 through 10 and the Model Inventory Appendix have the advantage of being already written. The user provides input data without being concerned about formulating mathematical algorithms and writing code. Most of the packages provide flexible optional capabilities which the user selects through input data entries. In constructing models using STELLA, EXTEND, Lotus 1-2-3, Quattro Pro, or Excel, the user must develop and code the computational algorithms as well as provide the input data. The relative difficulty in applying the two alternative types of software tools depends largely on the complexity of the system being modeled and the computational methods. In general, STELLA, EXTEND, and spreadsheet programs are more amenable to simpler problems. For example, a water distribution system (Chapter 5) with 10 pipes and no pumps would be significantly easier to model with a spreadsheet program than a system with 500 pipes and several pumps. Likewise, a reservoir/river system operation model (Chapter 10) for a system with three water supply reservoirs is much easier to formulate with STELLA than a model for a system of 12 multiple-purpose (water supply, hydropower, and flood control) reservoirs. A finite difference solution of the St. Venant partial differential equations of flow is more difficult to code than a hydrologic streamflow routing technique (Chapter 8). From a user perspective, the generalized water management models are not nearly as sensitive to the size and complexity of the problem because the computational methodologies are already formulated and coded. STELLA, EXTEND, and the spreadsheet programs provide the advantages of being well-polished, user-oriented, flexible packages which can be routinely applied for many different tasks.

EXPERT SYSTEMS

Artificial intelligence (AI) is the field of computer science that deals with making machines behave in a way that would generally be accepted as requiring human intelligence. Subfields of artificial intelligence include robotics, computer vision, speech synthesis and recognition, automated reasoning and theorem proving, natural language processing, automatic programming, automated learning, neural networks, and expert systems. Although all of these subfields are of interest to researchers, only some have reached the stage of commercial applicability. Of these, expert systems is the subfield of artificial intelligence that has been most extensively applied in business, industry, and various professions (Prerau 1990).

An expert system is an advanced computer program that can, at a high level of competence, solve difficult problems requiring the use of expertise and experience (Prerau 1990). Expert systems may be stand-alone models or may be embedded as a component of a larger model. Expert systems employ knowledge of the techniques, information, heuristics (rules of thumb), and problem-solving processes used by human experts. Expert systems are often called knowledge-based systems. They provide a way to store human knowledge, expertise, and experience in computer systems. Expert systems serve the following purposes:

- Providing expertise at times and locations where and when experts are not available
- Freeing the time of experts for other more important tasks
- Preserving knowledge which might otherwise be lost through retirement, reassignment, or other loss of personnel
- Upgrading the performance of less experienced and less skilled personnel
- Minimizing resources required to train new personnel
- Ensuring that expertise is applied uniformly, objectively and consistently
- Automating expertise that can be marketed as a product or service

An expert system consists of three basic components: knowledge base, inference engine, and working memory. The knowledge base contains the methods, information, and heuristics used by experts in the particular application area or domain. It also includes expert techniques on how and when to use these facts and heuristics. The working memory contains the information the system has received and derived about the particular problem currently being addressed. The inference engine provides the system control. It applies the expert domain knowledge from the knowledge base to what is known about the present situation (as reflected by the information currently in the working memory) to solve the problem or make a decision. The information in the knowledge base and working memory, which is manipulated by the inference engine, is typically made up primarily of symbols representing concepts, with relatively little numerical data.

Expert System Software

Two alternative types of software tools are available for developing expert systems: programming languages and shells. The two most widely used AI languages are LISP and PROLOG (Hopgood 1993). Languages such as LISP and PROLOG are used to develop expert systems just as FORTRAN and BASIC are used for conventional programming. LISP (LISt Processor) is a traditional AI language that dates back to the 1950s. LISP is designed to process primarily symbolic information rather than the numerical data associated with more familiar languages such as FORTRAN. PROLOG (PROgramming in LOGic) is another AI language designed to manipulate logical expressions. PROLOG was originally developed in Europe but is also widely used in the United States.

Expert system shells represent an alternative to programming languages like LISP and PROLOG. A shell is essentially an expert system with an empty knowledge base (Hopgood 1993). An expert system can be developed by purchasing a shell and building a knowledge base. Many expert system shells, of varying complexity, are marketed by various software companies. A number of different shells have been adopted in water management applications, with none being more popular than the others. A shell includes an inference engine, a user interface for programming, and a user interface for running the system. Shells are designed to simplify development of expert systems. Shells essentially implement a subset of a language that is important for a particular type of application. Because all application domains are different, it is difficult for a software supplier to build a shell that adequately handles a broad range of applications. Therefore, although use of shells is typically simpler, writing programs using LISP or PROLOG provides greater flexibility.

Expert System Applications

During the past two decades, expert systems have been applied in a variety of fields, including medicine, manufacturing, engineering, law, meteorology, geology, and information management. In recent years, expert systems have received significant attention in environmental and water resources management. Hushon (1990) and Wright et al. (1993) review environmental applications of expert systems. Hushon (1990) includes a summary of a 1989 survey of 69 environmental expert systems. Several expert systems reported in the recent literature are cited here to illustrate the range of applications in water management.

A number of expert systems have been developed to guide wastewater treatment plant operators, including those reported by Stover and Campana (1991), Laukkanen and Pursiainen (1991), and Hale (1991). Nix and Collins (1991) describe one of several reported expert systems used to make operating decisions for water treatment plants.

Several expert systems have been developed at the USACE Waterways Experiment Station. ENDOW (Environmental Design of Waterways) is an expert system which aids planners and designers in selecting environmental features for stream channel alteration projects based on key project parameters and specific environmental goals (Miller 1992; Shields and Schaefer 1990). A knowledge-based system called PUMP was developed, using PROLOG, to guide reservoir operators in evaluating localized mixing alternatives for alleviating water quality problems associated with thermal stratification (Price et al. 1992). Various other entities also have developed expert systems for guiding reservoir operations. The Consequences of Bioaccumulation in Aquatic Animals (COBIAA) expert system was developed to help interpret bioaccumulation tests conducted in the regulatory evaluation of dredged material (Dillon and Lutz 1991).

The Cornell Mixing Zone Expert System (CORMIX), maintained by the Environmental Protection Agency Center for Exposure Assessment Modeling, is used for the analysis, prediction, and design of aqueous toxic or conventional pollutant discharges into watercourses, with emphasis on the geometry and dilution characteristics of the initial mixing zone (Jirka et al. 1991). Males et al. (1992)

investigated the use of expert system technology to support personnel of a water utility in Pennsylvania in handling customer complaints. Engel and Beasley (1991) developed a Dam Site Selector (DSS) expert system as an aid in evaluating alternative sites for construction of dam and reservoir projects.

Several expert systems are available for developing parameter values for water management models. For example, Barnwell and Brown (1989) describe an expert advisor for guiding model users in developing input data for the QUAL2E stream water quality model discussed in Chapter 9. Baffaut and Delleur (1990) and Liong et al. (1991) have developed knowledge-based systems for calibrating the Stormwater Management Model (SWMM) discussed in Chapter 7. McClymont and Schwartz (1991) describe an expert system to guide input data preparation for a ground-water contaminant transport model.

EQUATION SOLVERS AND MATHEMATICAL MODELING ENVIRONMENTS

Water management models are based on sets of algebraic or differential equations representing governing principles such as conservation of mass, momentum, or energy. For fairly simple applications, model development may consist of formulating the appropriate equations and then solving them using mathematics programs. Mathematica, MathCAD, MATLAB, and TK! Solver are listed in Table 3.3 as examples of the many available equation solvers or mathematical model building

TABLE 3.3   Equation Solvers and Mathematical
                Modeling Environments

MathCAD	MathSoft, Inc.
	201 Broadway Street
	Cambridge, MA 02139
	(800)628-4223, (617)577-1017
Mathematica	Wolfram Research, Inc.
	100 Trade Center Drive
	Champaign, IL 61820-7237
	(800)441-MATH, (217)398-0700
MATLAB	The MathWorks, Inc.
	24 Prime Park Way
	Natick, MA 01760
	(508)653-1415
TK! Solver	Universal Technical Systems, Inc.
	1220 Rock Street
	Rockford, IL 61101
	(800)435-7887, (815)963-2220

environments. These software products provide capabilities for solving algebraic equations; performing differentiation and integration, matrix operations, and statistical computations; and displaying the results in numbers, tables, symbols, and graphs. Many built-in statistical, transcendental, and other math functions are provided. The interactive programs are designed to be user friendly with nice graphics. The examples cited and other similar programs are priced at several hundred dollars and are available for IBM-compatible microcomputers operating under MS-DOS as well as for other common computer systems.

MATHEMATICAL PROGRAMMING (OPTIMIZATION) MODELS

Mathematical programming or optimization involves determining values for a set of decision variables which will minimize or maximize an objective function subject to a set of constraints. Linear programming involves optimizing a linear objective function subject to a set of linear constraints. A variety of nonlinear programming methods are available for solving problems with nonlinear terms. Practical problems often involve hundreds or perhaps many thousands of decision variables and constraint equations.

Mathematical programming is widely applied in water resources planning and management as well as in many other fields. Optimization models are a central focus of the various systems engineering fields, including water resources systems engineering. Thousands of journal and conference papers and other publications have reported research in applying linear programming, dynamic programming, and other nonlinear programming techniques to various types of water resources problems. As discussed in Chapter 10, a particularly large number of reservoir system optimization models have been reported in the literature. Yeh (1985) presents a state-of-the-art review of the use of mathematical programming methods in analyzing reservoir system operations. Yeh (1992) reviews the application of optimization techniques in ground-water modeling, particularly in regard to the inverse problem of calibrating parameters for simulation models. Goulter (1992) reviews the use of optimization models in designing components of water distribution systems. Agricultural economics models for forecasting irrigation water demands are also often based on linear programming.

Water management optimization models are often written from scratch in FORTRAN or other languages for a particular application. Already-written FORTRAN subroutines for performing linear or nonlinear programming computations are often incorporated into the coding of water management optimization models. The same code for the optimizer routines may be used in any number of different models. These subroutines to be embedded within FORTRAN programs would not necessarily be considered generalized operational models. Another modeling ap-

TABLE 3.4   Mathematical Programming
(Optimization) Software

GAMS	Scientific Press
	651 Gateway Blvd., Ste. 1100
	South San Francisco, CA 94080
	(800)451-5409, (415)583-8840
Lindo	Lindo Systems, Inc.
	1415 N. Dayton Street
	Chicago, IL 60622
	(800)441-2378, (312)871-2524
LP88	Eastern Software Products, Inc.
	P.O. Box 15328
	Alexandria, VA 22309
	(703)360-7600, (703)360-7654

proach is to use a generalized mathematical programming package to create the water management model without performing any FORTRAN programming at all.

Recent versions of the previously discussed spreadsheet programs (Table 3.1) include linear programming capabilities but are not designed for solving problems with extremely large numbers of variables and constraints. The mathematical programming software listed in Table 3.4 is representative of the various available programs which are designed specifically for solving linear and, in some cases, nonlinear programming problems, including very large problems. LP88 is a linear programming code. Lindo solves linear, integer, and quadratic programming problems. The user inputs values for the coefficients in the objective function and constraint equations for the problem formulation of concern. The optimizer program computes values for the decision variables.

The General Algebraic Modeling System (GAMS) is a general-purpose optimization package designed for solving large linear, nonlinear, and mixed integer programming problems (Brooke et al. 1992). GAMS is a high-level language that provides data management and model formulation capabilities as well as a set of mathematical programming optimizers. GAMS was originally developed by the World Bank (International Bank for Reconstruction and Development) of the United Nations. The GAMS system consists of GAMS/MINOS and GAMS/ZOOM in addition to the basic GAMS module. GAMS/MINOS (Modular In-Core Nonlinear Optimization System) accepts GAMS-formulated input to solve complex linear and nonlinear programming problems. GAMS/ZOOM (Zero/One Optimization Method) accepts GAMS-formulated input to solve mixed integer programming problems. GAMS also includes a library of linear and nonlinear programming models that have been formulated for various applications. GAMS is available in microcomputer, workstation, and mainframe versions.

Labadie (1990) describes a generalized microcomputer model for solving dynamic programming problems called CSUDP, which was developed at Colorado State University specifically for water resources planning and management applications. CSUDP has been used to develop a variety of different types of water management models.

STATISTICAL ANALYSIS PROGRAMS

Statistical analyses have broad applicability in water resources planning and management. Many of the water management models discussed in Chapters 4 through 10 contain built-in statistical analysis capabilities. Most of the software packages cited in this chapter include some capabilities for statistical analysis of data. Commercial statistics packages providing comprehensive capabilities for data analysis are also useful in water management modeling studies. The statistical analyses may involve either input or output data for the models addressed in Chapters 4 through 10. Several general-purpose proprietary statistics programs are listed in Table 3.5 as examples of the many general statistical analysis packages on the market that provide capabilities for database management, graphics, and report generation along with a variety of statistical computations (including descriptive statistics, various types of statistical tests, regression and correlation analyses, and time series analyses). Several public domain statistical analysis packages developed specifically for water resources applications are also noted in the following discussion.

TABLE 3.5 Statistical Analysis Programs

BMDP/PC	BMDP Statistical Software, Inc.
	1440 Sepulveda Blvd., Ste., 316
	Los Angeles, CA 90025
	(800)238-2637, (310)479-7799
SAS/ETS	SAS Institute, Inc.
	100 SAS Campus Drive
	Cary, NC 27513-2414
	(919)677-8000
SPSS/PC+	SPSS, Inc.
	444 N. Michigan Avenue
	Chicago, IL 60611-3962
	(800)543-9262, (312)329-2400
StatGraphics	Statistical Graphics Corp.
	Five Independence Way
	Princeton, NJ 08540
	(800)232-STAT, (609)924-9374

Public domain models available from the USACE Hydrologic Engineering Center (Tables 2.1 and 2.2) include several statistical analysis programs. The Statistical Analysis of Time Series (STATS) program is designed to reduce large volumes of time series data to a few meaningful statistics or curves. The program will perform the following analyses: (1) duration curves, (2) annual minimum and maximum events, (3) departures of monthly and annual values from respective means, and (4) annual volume-duration exchange of high and low events. Data may be provided directly to STATS as an input file or accessed through the HEC Data Storage System (HEC-DSS), which is described later in this chapter. STATS can access output data from various simulation models as well as other databases through HEC-DSS. Frequency analyses may be either analytical or graphical. The Flood Frequency Analysis (HEC-FFA) program performs frequency analyses of annual flood peaks in accordance with a standard procedure outlined by the Water Resources Council Bulletin 17B. The Regional Frequency (REGFRQ) program computes statistics of annual maximum hydrologic events for use in regional frequency studies. The Multiple Linear Regression Program (MLRP) includes features for automatic deletion of independent variables according to importance; combination of variables to form new variables; transformation of variables; tabulation of the residuals from the prediction equation; and acceptance of input coefficients. The HEC-4 Monthly Streamflow Simulation program may be used to analyze monthly flows at a number of interrelated stations to determine their statistical characteristics and generate sequences of hypothetical flows of any desired length based on those statistics. HEC-4 will also reconstitute missing flows on the basis of concurrent flows observed at other locations. Executable versions of STATS, HEC-FFA, and MLRP are available for MS-DOS microcomputers. FORTRAN source codes are available for HEC-4 and REGFRQ.

The general-purpose statistics programs listed in Table 3.5 include capabilities for various types of time series analyses. Models have also been developed specifically for dealing with hydrologic time series. The LAST (Lane and Frevert 1990) and Spigot (Grygier and Stedinger 1990) as well as HEC-4 provide capabilities for analysis and synthesis of streamflow data. The statistical characteristics of historical flows, including relationships between flows at different locations, are analyzed. Sequences of hypothetical flows are then generated which preserve selected statistical characteristics of the historical flows.

Geostatistics deals with statistical analysis of spatially distributed quantities. It involves managing large data sets of spatially distributed samples and extensive computations. Geostatistics is important in ground-water modeling (Chapter 6) for analyzing spatial data such as head, layer thickness, hydraulic conductivity, storativity, temperature, and concentration. Kriging programs are commonly used to estimate variable values at model grid notes based on observed values at scattered and irregular monitoring sites.

GEOPACK (Yates and Yates 1990) is a public domain software system available from the EPA Robert S. Kerr Environmental Research Laboratory and the International Ground Water Modeling Center (Table 2.1). GEOPACK is a package of programs for statistical analysis of spatially correlated data. A database with up to 10,000 storage locations and containing a maximum of 10 variables plus their x and y positions and a sample or position number is supported. Statistical programs determine the mean, median, variance, standard deviation, skew, kurtosis, and maximum and minimum values. Linear and polynomial regression options are provided. The Kolomogorov-Smirnov test for probability distribution can be performed. Percentiles of a data set are also computed. Variogram programs allow determination of a sample semivariogram, the cross-semivariogram, or a combined random functions semivariogram. GEOPACK also can compute ordinary kriging and cokriging estimators in two dimensions and their associated estimation variance. Nonlinear estimators such as the disjunctive kriging and cokriging estimators can also be determined. GEOPACK includes various graphics capabilities such as linear or logarithmic line plots, contour plots, and block diagrams. GEOPACK output files can also be exported to commercial graphics software.

GEO-EAS (Englund 1990), also available from the International Ground Water Modeling Center, is a collection of interactive software tools for performing geostatistical analysis of spatially distributed data. Programs are provided for data file management, data transformations, univariate statistics, variogram analysis, cross validation, kriging, contour mapping, postplots, and line/scatter graphs. Both GEO-EAS and GEOPACK are available in for MS-DOS-based microcomputers.

DATA MANAGEMENT SYSTEMS

Water management models often require voluminous sets of hydrologic, climatic, water quality, and other types of input data. Simulation results typically are voluminous. Many of the water management models cited in Chapters 4 through 10 have pre- and/or postprocessor programs, developed for the individual models, that are used to store, transport, organize, manipulate, analyze, summarize, and display model input and output data. Other data management and analysis software is not associated with one particular simulation model but rather can serve as a preprocessor for input data and/or postprocessor for output data for a variety of different models. Commercially available database management systems are widely applied in business, engineering, and other fields and can also be used for water management purposes. Several public domain data management systems have been developed by federal agencies and others specifically for water resources applications.

Commercial Database Management Programs

The previously discussed spreadsheet/graphics/database management packages, such as Lotus 1-2-3, Quattro Pro, and Excel, are sometimes used as pre- and postprocessors of input and output data associated with water management models. Other database management systems are available which do not have the spreadsheet computational features but provide expanded database management capabilities.

Database management systems are widely used in government, business, industry, and education to manage a diversity of types of information. A database management system is a collection of numerous interrelated software routines. The primary functions of a database management system are to

- Create and organize the database
- Set up and maintain access paths so that data in any part of the database can be accessed quickly
- Manipulate the data in response to user requests
- Maintain data integrity and security
- Perform logging on database use (Tsai 1988)

Database management systems provide storage and access capabilities for large amounts of data. The same data can be used for many different types of applications programs.

Numerous database management systems are available for mainframe computers, minicomputers, and microcomputers. Paradox (currently version 4), dBASE (versions dBASE II, III, and IV), and Access are popular programs for MS-DOS-based microcomputers. Versions are available for other microcomputer systems as well. Both Paradox and dBASE are available from Borland International, which is listed in Table 3.1 with Quattro Pro. Both are priced at about $800. dBASE was originally developed by the Ashton-Tate Company, which was subsequently acquired by Borland International. dBASE was one of the earliest and is still one of the most widely used database management programs. Paradox is also popular, has capabilities which are similar to dBASE, but is somewhat simpler and easier to use than dBASE. Access is a similar newer database package marketed by Microsoft. The programs are used primarily for business administration applications but also in engineering and other areas, including water management.

Paradox, dBASE, Access, and similar programs provide a library of commands for defining the database file structures, loading the files, and manipulating data in the databases. The user can write programs to perform various tasks by combining commands. The user works within an interactive menu-driven environment. The systems are said to be relational because each database file is considered as a two-dimensional table, and related files are linked via connection fields.

Water Resources Data Management and Analysis Programs

A number of software packages have been developed by the water agencies specifically for water resources-related data management and analysis applications. Two such packages, HECDSS (USACE Hydrologic Engineering Center 1990) and ANNIE (Lumb et al. 1990), are discussed next and can be used, in conjunction with various water management models, to perform a broad range of tasks associated with preprocessing of input data and postprocessing of model results.

HECDSS.   The Hydrologic Engineering Center (HEC) Data Storage System (DSS) is available from the U.S. Army Corps of Engineers Hydrologic Engineering Center. The HEC Data Storage System (HECDSS) is used routinely with several HEC simulation models, including HEC-1, HEC-5, HECPRM, and WQRRS, discussed in subsequent chapters. HECDSS can be used with other non-HEC programs as well. The database could include any type of data but would typically consist of hydrologic, climatic, water quality, or economic data. HECDSS database management capabilities are oriented particularly toward voluminous sets of time series data. HECDSS uses a block of sequential data as the basic unit of storage. The basic concept underlying the HECDSS is the organization of data into records of continuous, applications-related elements as opposed to individually addressable data items. This approach is more efficient for water resources applications than that of a conventional database system because it avoids the processing and storage overhead required to assemble an equivalent record from a conventional system.

HECDSS is available for desktop computers running the MS-DOS operating system. The software package has also been compiled and executed on various minicomputer and mainframe systems. The programs are written in FORTRAN77.

HECDSS provides capabilities to (1) store and maintain data in a centralized location, (2) provide input to and store output from application programs, (3) transfer data between application programs, (4) mathematically manipulate data, and (5) display the data in graphs and tables. The user may interact with the database through (1) utilities that allow entry, editing, and display of information; (2) application programs that read from and write to the database; and (3) library routines that can be incorporated in any program to access database information.

A variety of utility programs are included in HECDSS for entering data into a database file. Some are designed for entering data from other databases, such as the U.S. Geological Survey WATSTORE system. Several HEC application programs have been interfaced with DSS, allowing users to retrieve data for analysis or store results in a DSS file. This provides the user the capability of displaying and analyzing application program results by using the DSS utility programs. A set of FORTRAN subroutines are available which can be used to link application programs with HECDSS.

HECDSS also provides means for mathematically manipulating data in a variety of ways. Normal arithmetic operations and many mathematical functions are provided. Various statistical analyses can be performed. Missing data can be synthesized. Hydrologic routing of streamflows can be performed.

ANNIE.   ANNIE is a computer program for interactive hydrologic analyses and data management which was developed by the U.S. Geological Survey (Lumb et al. 1990). ANNIE contains a set of procedures to organize, manipulate, and analyze data needed for hydrologic modeling and analysis. The user interactively performs tasks related to data management, tabular and graphical presentation, statistical analysis, and input preparation for hydrologic models. ANNIE stores data in a binary, direct-access file with a specified structure, which is called a Watershed Data Management (WDM) file. The WDM file provides users with a common database for many applications, thus eliminating the need to reformat data from one application to another. ANNIE and/or the WRM file format are currently used with a number of U.S. Geological Survey and Environmental Protection Agency hydrologic and water quality models. ANNIE is written in FORTRAN and is designed for portability to mainframe computers, minicomputers, and microcomputers.

ANNIE provides capabilities to create a WDM file conveniently, transport data to and from the WDM file, and adjust and manipulate the data. The data can be tabulated in various presentation formats. ANNIE graphics capabilities include time series plots, x-y plots, and probability plots. The plots can meet USGS publication standards. ANNIE provides a number of statistical analysis capabilities, including flow-duration, frequency, error, and trend analyses. The ANNIE library of routines has also been used to create custom programs for use in developing input files for specific hydrologic, hydraulic, or water quality simulation programs. Some custom application input data development programs are available, and users can write their own programs for other applications.

ANNIE-IDE.   The ANNIE Interaction Development Environment (ANNIE-IDE) was created by the Environmental Protection Agency to provide a consistent methodology for building interactive interfaces for environmental computer programs and databases (Kittle et al. 1989). ANNIE-IDE incorporates a number of routines and methods from ANNIE. ANNIE-IDE is a set of tools for developing user interfaces for simulation models and pre- and postprocessor programs. The ANNIE-IDE system provides the program developer with a set of subroutines which may be incorporated into a model to perform one or more of the following operations: (1) display text on the monitor screen, (2) display static and/or dynamic menus, (3) prompt the user to input or edit values in a one- or two-dimensional array, (4) open a file to store or retrieve information, and (5) display context-sensitive help, instructions, and model parameter information.

GEOGRAPHIC INFORMATION SYSTEMS

A geographic information system (GIS) is a set of computer-based tools for storing, processing, combining, manipulating, analyzing, and displaying data which are spatially referenced to the earth. Thus, GIS is a special case of data management and analysis dealing specifically with spatial or geographical data. The development and application of GIS technology is outlined in detail by Antenucci et al. (1991) and Maquire et al. (1991). Goodchild et al. (1993) address the use of GIS in environmental modeling. GIS technology dates back to the 1960s and has evolved into a major discipline in recent years. Interest in GIS applications in water resources planning and management expanded greatly during the late 1980s and 1990s.

GIS Components and Capabilities

A geographic information system is comprised of people, computer hardware and software, and data. Trained personnel are required to design, implement, maintain, and use a GIS. A significant level of expertise is normally required to develop and apply a GIS. The lack of adequately trained personnel has been highlighted by Maquire et al. (1991) and others as a significant constraint, in some cases, to successful implementation of available technology.

The hardware component of the GIS can include almost any type of computer platform ranging from relatively modest microcomputers to high-performance work-stations and minicomputers to mainframe computers. Workstations running under the Unix operating system are particularly popular for GIS applications. GIS hardware typically includes a digitizer and plotter as well as the central processing unit and visual display unit.

Development and marketing of GIS software is a sizable industry. Numerous software packages providing a broad range of capabilities are available. Two popular software packages, ARC/INFO and GRASS, are discussed later.

The last, but not least, element of the GIS is data. GIS applications normally involve large volumes of data. Building a database can be expensive and time consuming. The types of data included in a GIS vary greatly depending on the particular applications. GIS data can be categorized as graphic and nongraphic. A GIS uses graphic data to generate maps or cartographic pictures. Graphic data are digital descriptions of map features. They may include the coordinates, rules, and symbols that define specific cartographic elements on a map. Nongraphic data are representations of the characteristics, qualities, or relationships of map features and geographic locations. Nongraphic data are often called textual data or attributes.

Graphic images are stored either as vectors or as a raster of uniform grid cells or pixels. Vector data are represented by horizontal x-y coordinates of point and line locations or as rules for computing the coordinates and connecting the points as lines or areas. Vector data define polygons, objects, and other complex entities that can be manipulated or displayed on the basis of their attributes. The alternative raster approach of representing data is based on uniform grid cells of specified resolution. A value for an attribute is assigned to each grid cell. Thus, data are sorted spatially as values for a matrix of grid cells.

Graphic data are often viewed as a series of layers. The database for a GIS for an urban area might include layers representing topographic contours, soils, streets, utilities, land ownership parcels, zoning districts, and municipal boundaries. For the soils layer, nongraphic attributes might include soil type, moisture content, and erosion parameters. For the utility layer, nongraphic attributes might include the size, material, and year installed for each water line. For a land ownership parcel, nongraphic attributes might include owner name, parcel size, land use, and market value of property.

Uses of GIS technology vary widely. In general, geographic information systems provide cartographic, data management, analytical, and polygon processing capabilities. Cartographic capability allows accurate maps and engineering drawings to be produced efficiently. This capability includes digitizing, graphic display generation, interactive graphic manipulation, and plotting. Data management capabilities involve the efficient storage and retrieval of both graphic and nongraphic data, including nongraphic attributes linked to graphic images. Data management includes selecting data and producing graphics and reports on the basis of attribute values. Analytical capabilities involve various mathematical computations and analyses. Polygon processing consists of overlaying sets of data. For example, soil type data may be overlain on land use data to construct polygons that have specified combinations of soil type and land use.

GIS Software

In many cases, customized GIS software has been developed specifically for a particular governmental organization or private company and its own particular applications. However, many generalized GIS software packages are available which provide a variety of capabilities for a broad range of applications. Two popular software packages, ARC/INFO and GRASS, are discussed next. ARC/INFO is a proprietary software package developed and marketed by Environmental Systems Research Institute, Inc. GRASS is public domain software developed and distributed

TABLE 3.6   GIS Software Packages

ARC/INFO	Environmental Systems Research Institute
	380 New York Street
	Redlands, CA 92373
	(909)793-2853
GRASS	Environmental Division
	Construction Engineering Research
	Laboratory
	U.S. Army Corps of Engineers
	Champaign, IL 61826
	(217)352-6511

by the Construction Engineering Research Laboratory of the U.S. Army Corps of Engineers. The GIS computer programs, documentation, reference materials, and information regarding training opportunities are available by contacting the developers at the addresses and telephone numbers provided in Table 3.6.

ARC/INFO.   ARC/INFO is one of the earliest and still most widely used of the available generalized GIS software packages. Early versions of ARC/INFO were developed for use on Sun workstations with the Unix operating system, but versions are currently available for implementation on a wide range of computer systems, including MS-DOS-based IBM-compatible microcomputers, Macintosh microcomputers, essentially all available workstations, and various mainframe computers. Significant effort is required to become proficient with the software. An array of manuals, references, and training materials is available to assist users.

As the name suggests, ARC/INFO is comprised of two components, ARC and INFO. ARC is a system for working with map coordinate data representing geographic features. INFO is database management system for attribute data. ARC/INFO is a set of tools for creating, analyzing, displaying, and managing computerized maps in vector format. The vector approach for storing data is used. Geometric features of a map are represented by points, lines (an arc or set of arcs), and polygons (planes enclosed by arcs). For example, a river basin application might involve representation of precipitation and streamflow gages as points, streams as arcs, subwatershed boundaries as arcs, and subwatersheds as polygons. The associated attribute data might includes stream reach lengths, subwatershed areas, soil types, and land use.

The ARC/INFO system provides a broad range of optional capabilities for data management and analysis. Data can be edited, checked, and manipulated in various ways. Data can be displayed in a variety of graphic and textual formats. Analytical tools are available for modeling networks, including the computational tasks of routing, allocating, and districting. Routing determines the optimum paths for the movement of resources (such as vehicles, water, electricity, or pulses of communication) through a network (comprised of roads, pipes, or telephone lines). Allocating involves finding the nearest center for each link in the network that best serves the network (such as finding the closest fire station from each street within a city). Districting involves aggregation of areas bounded by certain networks, such as dividing a city into districts bounded by selected streets. Attribute data can then be displayed by district.

GRASS. The Geographical Resources Analysis Support System (GRASS) was originally developed by the USACE Construction Research Laboratory in the early 1980s and has since evolved through various versions and expansions. GRASS is a public domain software package. The source code, written in the C language, is available to users. Improvements and additions to the software by various users are incorporated as new versions are released. The GRASS user community is large and includes GIS software vendors as well as end-users. The Open GRASS Foundation (Champaign, IL 61826-3879, 217/352-6511) is a nonprofit organization created to coordinate the activities of the user community in the continued development and application of GRASS. The Open GRASS Foundation continues the work of the earlier GRASS Inter-Agency Steering Committee and GRASS GIS User Forum. *GRASSCLIPPINGS* was published by the Open GRASS Foundation for several years as a newsletter and was converted to a journal in 1992. The European GRASS Foundation publishes the *European GRASS Community Newsletter* and encourages interactions among users in Europe.

GRASS is a raster (grid cell)-based GIS as contrasted with ARC/INFO, which is vector based. GRASS uses the raster format for all its image and geographical data processing. In a raster format, the landscape is divided into a grid of rectangular parcels of land. Associated with each parcel are characteristics or attributes like land use, vegetation, geology, slope, elevation, etc. GRASS uses the vector form for human and machine input which has distinct lines, such as counties, cities, reservoirs, and watersheds. However, the vector data must be translated into raster format before performing any type of analysis.

The GRASS package includes about 200 computer programs, which the user can run directly through commands at the keyboard or indirectly through menus and other programs. Capabilities are provided for geographical analysis, image processing, map display, and data input. Geographical analysis capabilities include proximity analyses, logical (and, or, not) reasoning, weighted overlays, and neighborhood processing. Image processing tools perform geographic referencing and classification tasks required to incorporate information from satellite images and high-altitude photography into the GIS database. Many display capabilities allow the user to generate landscape images and maps for computer monitor and hardcopy display. GRASS includes an array of programs which allow data to be input or read from paper maps, satellite images, other computers, and various other sources.

GIS Applications

Maquire et al. (1991) and Antenucci et al. (1991) describe applications of GIS technology by federal, state, and local governmental agencies and private companies in the fields of urban planning, transportation, utilities, land management, environmental resources management, forestry, mineral exploration, and military activities. Numerous GIS applications in water management have also been reported in the published literature, particularly during the past few years. Starr and Anderson (1991) and Kilgore et al. (1993) review the use of GIS technology by the U.S. Geological Survey and U.S. Army Corps of Engineers, respectively. The Universities Council on Water Resources (1992) devoted a recent issue of its *Water Resources Update* to review GIS applications in water resources. Goodchild et al. (1993) also outline applications of GIS in water management modeling. The several applications cited next illustrate the broad range of uses of GIS in water management.

Mapping of water distribution and wastewater collection systems and other water-related facilities is a major application of GIS. For example, in 1990, the City of New York began a five-year mapping project, using GIS, to compile a digital large-scale base map of the city's 6,000 miles of water mains—which include some 180,000 valves, 99,000 fire hydrants, and numerous other appurtenances (Moutal and Bowen 1991). Monical (1992) describes a statewide GIS of rural water delivery systems in Arkansas.

A number of applications of GIS in watershed modeling are reported in the literature. For example, DeBarry and Carrington (1990) used ARC/INFO to manage watershed data in support of stormwater management planning for an urbanizing area in Pennsylvania. The data, which included topography, subwatershed boundaries, land use, and soil groups, were used to develop parameter values for a watershed model. Flood-plain management studies conducted by Skipwith et al. (1990) for a watershed in Dallas incorporated a GIS with remote sensing, global-positioning satellite surveying system, three-dimensional digital mapping and topography, and computer-aided drafting and design. The study analyzed properties and structures subject to flood damage under specified scenarios. Pearson and Wheaton (1993) describe the use of ARC/INFO by the City of Anchorage, Alaska in the application process for a National Pollutant Discharge Elimination System (NPDES) stormwater permit. The information managed by this GIS included parcel and base maps; land use; topography; population growth; rainfall data; streams, lakes, and wetlands; pipes and drainage systems; and soils and hydrologic data.

A number of studies of the impacts on water quality of human activities in watersheds have involved GIS applications. In implementing a program for water quality protection of supply sources, the Massachusetts Water Resources Authority used a GIS for data management and mapping in the delineation of ground-water and surface-water protection zones (Cherin and Brandon 1992). McCreary et al. (1992) describe the application of GIS in a study of the impact of future growth in the San Francisco Bay-Delta Region on the estuary, wetlands, streams, and water quality. Rifai et al. (1993) used ARC/INFO in an investigation of nonpoint pollution sources as part of the Galveston Bay National Estuary Program in Texas. Halliday and Wolfe (1991) used the GRASS GIS to correlate the availability of nitrogen fertilizer with the susceptibility of ground-water pollution in Texas. An agricultural pollution susceptibility map developed by the Texas Water Commission was combined with information on cropped areas, recommended nitrogen fertilizer application rates, and aquifer outcrops to produce a nitrogen fertilizer pollution potential index.

Haefner (1992) used GIS to compile and analyze geohydrologic, geologic, chemical, and other data required to evaluate the physical factors that determine the suitability of sites for public water supply wells. Orzol and McGrath (1992) document a computer program that interfaces between a GIS and the MODFLOW ground-water model.

Sinclair et al. (1990) describe the use of GIS and satellite images by the USACE Waterways Experiment Station in the mapping and analysis of wetland environments in the Mississippi River Valley. GIS technology has also been applied in a number of other wetlands studies reported in the literature.

COMPUTER-AIDED DRAFTING AND DESIGN

Computer-aided drafting and design (CADD) software is used for a variety of graphics applications in various fields, including water resources planning and management. CADD programs provide powerful drawing and specialized graphics capabilities. These programs are used for mapping and general technical illustration purposes as well as traditional drafting and design functions. Because CADD programs store and display spatial data, they often serve as components of geographic information systems. CADD programs are often used to provide drawing capabilities for the water management models described in Chapters 4 through 10. For example, the KYPIPE2 water distribution system program, discussed in Chapter 5, interchanges files with AutoCAD (AutoCAD is used to draw complex pipe networks).

AutoCAD is the most popular of the various CADD packages. AutoCAD is a general-purpose two- and three-dimensional drafting system employed in a broad

TABLE 3.7   Computer-Aided Drafting and Design
Software

AutoCAD and Generic CADD	Autodesk, Inc. 2320 Marinship Way Sausalito, CA 94965 (800)964-6433, (415)332-2344
MicroStation	Intergraph Corporation Huntsville, AL 35894-0001 (800)345-4856, (205)730-2000

range of applications. Generic CADD is a popular two-dimensional drawing system that does not provide all the capabilities of AutoCAD but is less expensive. MS-DOS microcomputer versions of AutoCAD (Release 12) and Generic CADD (Version 6.1) sell for $3,750 and $495, respectively. As indicated in Table 3.7, both AutoCAD and Generic CADD are marketed by Autodesk, Inc. MicroStation (Version 5), from the Intergraph Corporation, with a list price of $3,790, also provides comprehensive two- and three-dimensional drafting and design capabilities. In addition to being a stand-alone CADD product, MicroStation is the graphics nucleus for many other applications software packages available from the Intergraph Corporation.

GRAPHICS PROGRAMS

Essentially all of the software cited in this chapter includes graphics capabilities. Many of the water management models discussed in Chapters 4 through 10 include built-in graphics. Graphics programs developed for specific water management modeling systems may often be used with other models as well. Commercial graphics packages are also sometimes used in water resources planning and management applications. The results of a simulation model may be transported to a graphics program for analysis, interpretation, and presentation. Graphics programs may also be useful for analyzing input data for a water management model.

Numerous software products on the market focus specifically on creating graphs and charts. Several popular presentation graphics programs, such as Harvard Graphics (Software Publishing Corp., Santa Clara, CA) and Stanford Graphics (3-D Visions Corp., Torrance, CA), provide capabilities for preparing elaborate graphs for reports and formal presentations as well as technical data analysis. Illustration programs like CorelDRAW (Corel Corp., Ottawa, Canada) allow creation of artistic drawings and diagrams and provide clip art images and symbols. Other graphics programs are designed to plot voluminous data, such as that often produced by water management models. *PC Magazine* (Jones 1991) reviewed 12 graphics packages designed to produce high-quality graphs and charts for scientific and engineering applications. The 12 graphics programs are listed in Table 3.8 along with the companies that market them. Prices for these programs range from $130 to $500. Cricket Graph, Grapher, Grafit, and Tech*Graph*Pad develop two-dimensional graphs. Slidewrite Plus is a presentation graphics package with drawing tools and business-type charting as well as two-dimensional plotting capabilities. ProPlot and EasyPlot provide limited three-dimensional plotting capabilities in addition to two-dimensional capabilities. SDS and Surfer are primarily for plotting three-dimensional graphs. Axum, Graftool, and SigmaPlot are comprehensive graphics packages with a variety of two-dimensional plots, full range of charts, mathematical analysis capabilities, and data transformation functions. Axum and Graftool also plot three-dimensional graphs. All of these programs support at least 2,000 data points per

TABLE 3.8 Graphics Programs

Grapher	Surfer
Golden Software, Inc.	Golden Software, Inc.
809 14th Street	809 14th Street
Golden, CO 80401-1866	Golden, CO 80401-1866
(800)972-1021, (303)279-1021	(800)972-1021, (303)279-1021
Cricket Graph	ProPlot
Computer Associates International, Inc.	Cogent Software
One Computer Associates Plaza	1030 Pine Street
Islandia, NY 11788-7000	Menlo Park, CA 94025-3405
(800)CALL-CAI, (516)342-5224	(415)324-4360
EasyPlot	Tech*Graph*Pad
Spiral Software	Binary Engineering
15 Auburn Place	100 Fifth Avenue
Brookline, MA 02146	Waltham, MA 02154
(800)833-1511, (617)739-1511	(617)890-1812
SlideWrite Plus	SDS
Advanced Graphics Software, Inc.	Datanalysis
5825 Avenida Encinas, Ste. 105	P.O. Box 45818
Carlsbad, CA 92008	Seattle, WA 98145
(619)931-1919	(206)682-1772
Axum	Graftool
TriMetrix Inc.	3-D Visions Corporation
444 N.E. Ravenna Blvd. #210-ML	2780 Skypark Drive
Seattle, WA 98115	Torrance, CA 90505
(800)548-5653, (206)527-1801	(800)729-4723, (213)540-8818
Sigma Plot	GraFit
Jandel Scientific	Erithacus Software, Ltd.
2591 Kerner Blvd.	P.O. Box 35
San Rafael, CA 94901	Staines, Middlesex, TW18 2TG
(800)874-1888, (415)453-6700	United Kingdom
	(784)463-4672

chart, and most can handle many more. All but the three-dimensional packages Surfer and SDS provide several curve fitting algorithms. All will run on IBM-compatible microcomputers under MS-DOS. Most have versions for other computer systems as well.

Grapher and Surfer are widely used in the water resources field. Both are interactive menu driven. Grapher plots x-y graphs, with up to 32,000 data pairs, with up to 10 curves per graph, with arithmetic or logarithmic axes. It has six different types of curve fitting algorithms. Surfer generates contour maps and three-dimensional surface plots. It generates a spreadsheet database from the keyboard or imports ASCII data files. Inverse distance, minimum curvature, or kriging options are provided for interpolation between irregularly and scattered data points. Surfer is used particularly in ground-water modeling. An example of a common use for Surfer is to plot the water table or piezometric surface generated by ground-water models.

REFERENCES

ANTENUCCI, J. C., K. BROWN, P. L. CROSWELL, and M. J. KEVANY, *Geographic Information Systems, a Guide to the Technology*, Van Nostrand Reinhold, 1991.

BAFFAUT, C., and J. W. DELLEUR, "Calibration of SWMM Runoff Quality Model with Expert System," *Journal of Water Resources Planning and Management*, American Society of Civil Engineers, Vol. 116, No. 2, March/April 1990.

BAKER, K. A., M. S. FENNESSY, and W. J. MITSCH, "Designing Wetlands for Coal Mine Drainage: An Ecologic-Economic Modelling Approach," *Ecological Economics*, Vol. 3, No. 1, March 1991.

BARNETT-MOORE, J. A., W. G. STANDARD, and B. D. FOSTER, "Utility Financial Planning and Rate Design by Computer," *Public Works*, Vol. 121, No. 11, October 1990.

BARNWELL, T. O., and L. C. BROWN, "Application of Expert Systems Technology in Water Quality Modeling" *Water Science and Technology*, International Association of Water Pollution Research and Control, Vol. 21, No. 8/9, London, U.K., 1989.

BOGEN, D. G., "Simulation Software for the Macintosh," *Science*, American Association for the Advancement of Science, Vol. 246, 1989.

BRADLEY, J. B., D. T. WILLIAMS, and M. BARCLAY, "Incipient Motion Criteria Defining Safe Zones for Salmon Spawning Habitat," *Proceedings of the 1991 National Hydraulic Engineering Conference*, American Society of Civil Engineers, 1991.

BROOKE, A., D. KENDRICK, and A. MEERAUS, *Release 2.25 GAMS, A User's Guide*, Scientific Press, South San Francisco, California, 1992.

CHERIN, P. R., and F. O. BRANDON, "Protecting Local Supplies: Perspectives From a Regional Water Purveyor," *Water Engineering and Management*, Vol. 139, No. 10, October 1992.

COSTANZA, R., F. H. SKLAR, and M. L. WHITE, "Modeling Coastal Landscape Dynamics," *BioScience,* American Institute of Biological Sciences, Vol. 40, No. 2, February 1990.

CRONK, J. K., W. J. MITSCH, and R. M. SYKES, "Effective Modelling of a Major Inland Oil Spill on the Ohio River," *Ecological Modeling,* Vol. 151, 1990.

DEBARRY, P. A., and J. T. CARRINGTON, "Computer Watersheds," *Civil Engineering,* American Society of Civil Engineers, Vol. 60, No. 7, July 1990.

DILLON, T. M., and C. H. LUTZ, "Computer-Assisted Expert System for Interpreting the Consequences of Bioaccumulation in Aquatic Animals (COBIAA)," *Environmental Effects of Dredging Information Exchange Bulletin,* Vol. D-91-4, USACE Waterways Experiment Station, Vicksburg, Mississippi, December 1991.

ENGEL, B. A., and D. B. BEASLEY, "DSS: Dam Site Selector Expert System for Education," *Journal of Irrigation and Drainage Engineering,* American Society of Civil Engineers, Vol. 117, No. 5, September/October 1991.

ENGLUND, E. J., "GEO-EAS Version 1.2," U.S. Environmental Protection Agency Monitoring System Laboratory, Las Vegas, Nevada, 1990.

FEISTUL, D. R., "Water Quality Modeling of a Chain of Lakes in a Rapidly-Developing Suburban Area Using the WERM Model," *Transferring Models to Users,* American Water Resources Association, Bethesda, Maryland, 1990.

GILES, J. E., R. K. JONES, P. A. MARCH, H Armour, and J. M. Epps, "Microcomputer-Aided Planning at a Hydro Control Center," *International Water Power and Dam Construction,* Vol. 42, No. 1, January 1990.

GOODCHILD, M. F., B. O. PARKS, and L. T. STEYAERT (editors), *Environmental Modeling with GIS,* Oxford University Press, 1993.

GOULTER, I. C., "Systems Analysis in Water-Distribution Design: From Theory to Practice," *Journal of Water Resources Planning and Management,* American Society of Civil Engineers, Vol. 118, No. 3, May/June 1992.

GRYGIER, J. C., and J. R. STEDINGER, "SPIGOT, A Generalized Streamflow Generation Software Package, Technical Description, Version 2.5," School of Civil and Environmental Engineering, Cornell University, Ithaca, New York, 1990.

HAEFNER, R. J., "Use of a Geographic Information System to Evaluate Potential Sites for Public-Water-Supply Wells on Long Island, New York," *Open File Report 91-182,* U.S. Geological Survey, Denver Colorado, 1992.

HALE, J. F., "Next Generation of Computerized Process Control," *Water Environment & Technology,* Vol. 3, No. 5, May 1991.

HALLIDAY, S. L., and M. L. WOLFE, "Assessing Ground Water Pollution Potential from Nitrogen Fertilizer Using a Geographic Information System," *Water Resources Bulletin,* American Water Resources Association, Vol. 27, No. 2, March/April 1991.

HANCOCK, M. C., and J. P. HEANEY, "Water Resources Analysis Using Electronic Spreadsheets," *Journal of Water Resources Planning and Management,* American Society of Civil Engineers, Vol. 113, No. 5, September 1987.

HIGHLAND, W. R., "Use of PC Spreadsheet Models as a Routine Analytical Tool for Solving Ground Water Problems," *Proceedings of NWWA Conference on Solving Ground Water Problems with Models,* National Water Well Association, Dublin, Ohio, 1987.

HOPGOOD, A. A., *Knowledge-Based Systems for Engineers and Scientists,* CRC Press, 1993.

HUSHON, J. M. (editor), *Expert Systems for Environmental Applications,* American Chemical Society, Washington, D.C., 1990.

JIRKA, G. H., R. L. DONEKER, and T. O. BARNWELL, "CORMIX: An Expert System for Mixing Zone Analysis," *Water Science and Technology,* Vol. 24, No. 6, 1991.

JONES, M., "Technical Graphics Packages," *PC Magazine,* Ziff Communications, Vol. 10, No. 6, March 26, 1991.

KARPACK, L. M., and R. N. PALMER, "The Use of Simulation Modeling to Evaluate the Effect of Regionalization on Water Supply Performance," Water Resources Series Technical Report No. 134, University of Washington, Seattle, Washington, March 1992.

KEYES, A. M., and R. N. PALMER, "The Role of Object Oriented Simulation Models in the Drought Preparedness Studies," *Water Management in the '90s, a Time for Innovation, Proceedings of 20th Annual WRPMD Conference,* American Society of Civil Engineers, New York, 1993.

KILGORE, R. T., J. S. KROLAK, and M. P. MISTICHELLI, "Integration Opportunities for Computer Models, Methods, and GIS Used in Corps Planning Studies," Report 93-R-5, USACE Institute for Water Resources, Fort Belvoir, Virginia, February 1993.

KITTLE, J. L., P. R. HUMMEL, and J. C. IMHOFF, "ANNIE-IDE, a System for Developing User Interfaces for Environmental Models, Programmers Guide," EPA/600/3-89/034, U.S. Environmental Research Laboratory, Environmental Research Laboratory, Athens, Georgia, April 1989.

LABADIE, J. W., "Dynamic Programming with the Microcomputer," *Encyclopedia of Microcomputers,* Marcel Dekker, New York, 1990.

LANE, W. L., and D. K. FREVERT, "Applied Stochastic Techniques, Personal Computer Version 5.2, User's Manual," Earth Sciences Division, U.S. Bureau of Reclamation, Denver, Colorado, 1990.

LAUKKANEN, R., and J. PURSIAINEN, "Rule-Based Expert Systems in the Control of Wastewater Treatment Systems," *Water Science and Technology,* Vol. 24, No. 6, 1991.

LIONG, S. Y., W. T. CHAN, and L. H. LUM, "Knowledge-Based System for SWMM Runoff Component Calibration," *Journal of Water Resources Planning and Management,* American Society of Civil Engineers, Vol. 117, No. 5, September/October 1991.

LOW, K., and C. ERICKSON, "Use of Interactive Simulation Environment to Model the Marais des Cynes-Osage River Basin," *Water Management in the '90s, a Time for Innovation, Proceedings of 20th Annual WRPMD Conference,* American Society of Civil Engineers, New York, 1993.

LUMB, A. M., J. L. KITTLE, and K. M. FLYNN, "Users Manual for Annie, A Computer Program for Interactive Hydrologic Analysis and Data Management," Water-Resources Investigations Report 89-4080, U.S. Geological Survey, Reston, Virginia, 1990.

LYNCH, C. J., "Demonstrating Competition for a Limited Resource in the Cedar/Green Basins," *Water Management in the '90s, a Time for Innovation, Proceedings of 20th Annual WRPMD Conference,* American Society of Civil Engineers, New York, 1993.

MACY, P. P., and W. O. MADDAUS, "Cost-Benefit Analysis of Conservation Programs," *Journal of the American Water Works Association,* Vol. 81, No. 3, March 1989.

MALES, R. M., J. A. COYLE, H. J. BORCHERS, B. G. Hertz, and W. M. Grayman, "Expert Systems Show Promise for Customer Inquiries," *Journal of the American Water Works Association,* Vol. 84, No. 2, February 1992.

MAQUIRE, D. J., M. F. GOODCHILD, and D. W. RHIND, *Geographical Information Systems: Principles and Applications,* two volumes, Longman Scientific & Technical, John Wiley & Sons, 1991.

McCLYMONT, G. L., and F. W. SCHWARTZ, "Embedded Knowledge in Software: 1. Description of a System for Contaminant Transport Modeling," *Ground Water,* Vol. 29, No. 5, September/October 1991.

McCREARY, S., R. TWISS, B. WARREN, C. WHITE, and S. HUSE, "Land Use Change and Impacts on the San Francisco Estuary: A Regional Assessment with National Policy Implications," *Coastal Management,* Vol. 20, No. 3, July/September 1992.

MEYER, E. B., "Pneumatic Destratefication System Design Using a Spreadsheet Program," *Information Exchange Bulletin,* Vol. E-91-1, U.S. Army Engineer Waterways Experiment Station, Vicksburg, Mississippi, March 1991.

MILES, S. W., and J. P. HEANEY, "Better than Optimal Method for Designing Drainage Systems," *Journal of Water Resources Planning and Management,* American Society of Civil Engineers, Vol. 114, No. 5, September 1988.

MILLER, J. L., "User Survey and Future of ENDOW," *Water Operations Technical Support Bulletin,* Vol. E-92-2, USACE Waterways Experiment Station, August 1992.

MONICAL, J. E., "User's Guide to the Arkansas Rural Water-Delivery Network Geographic Information (GIS) Software," Open-File Report 92-108, U.S. Geological Survey, Denver, Colorado, September 1992.

MOUTAL, H. P., and D. BOWEN, "Computer Mapping of New York City's Water Mains," *Water Engineering and Management,* Vol. 138, No. 10, October 1991.

NIX, S. J., and A. G. COLLINS, "Expert Systems in Water Treatment Operation," *Journal of the American Water Works Association,* Vol. 83, No. 2, February 1991.

NVULE, D. N., "Multiparty Model Development Using Object-Oriented Programming," *Water Management in the '90s, a Time for Innovation, Proceedings of 20th Annual WRPMD Conference,* American Society of Civil Engineers, New York, 1993.

OLSTHOORN, T. N., "The Power of the Electronic Worksheet: Modeling Without Special Programs," *Ground Water,* Vol. 23, No. 3, 1985.

ORZOL, L. L., and T. S. McGRATH, "Modifications of the U.S. Geological Survey Modular, Finite-Difference, Groundwater Flow Model to Read and Write Geographic Information System Files," Open-File Report 92-50, U.S. Geological Survey, Denver, Colorado, 1992.

PEARSON, M., and S. WHEATON, "GIS and Storm-Water Management," *Civil Engineering,* American Society of Civil Engineers, September 1993.

PRERAU, D. S., *Developing and Managing Expert Systems,* Addison-Wesley Publishing Company, 1990.

PRICE, R. E., J. P. HOLLAND, and S. C. WILHELMS, "Design of Localized Mechanical Mixing Systems," *Water Operations Technical Support Bulletin,* Vol. E-92-2, USACE Waterways Experiment Station, Vicksburg, Mississippi, August 1992.

RIFAI, H. S., C. J. NEWELL, and P. B. BEDIENT, "Getting to the Non-Point Source with GIS," *Civil Engineering,* American Society of Civil Engineers, June 1993.

SALGEONKER, J., "Planning Tool for Water Conservation," *Water: Laws and Management,* American Water Resources Association, Bethesda, Maryland, 1989.

SAMSTAG, R. W., A. J. LESTER, and F. S. THATCHER, "Automated Design of Wastewater Plants," *Civil Engineering,* American Society of Civil Engineers, Vol. 59, No. 6, June 1989.

SHATKIN, J. A., and H. S. BROWN, "Pharmacokinetics of the Dermal Route of Exposure to Volatile Organic Chemicals in Water: A Computer Simulation Model," *Environmental Research,* Vol. 56, No. 1, October 1991.

SHIELDS, F. D., and T. E. SCHAEFER, "ENDOW User's Guide," Instruction Report W-90-1, USACE Waterways Experiment Station, Vicksburg, Mississippi, December 1990.

SINCLAIR, R. H., M. R. GRAVES, and J. K. STOLL, "Satellite Data and Geographic Information Systems Technology Applications to Wetlands Mapping," Federal Coastal Wetland Mapping Programs, Fish and Wildlife Service Biological Report 90(18), December 1990.

SKIPWITH, W., L. MORLAND, M. DENTON, and M. ASKEW, "Closing the Floodgates," *Civil Engineering,* American Society of Civil Engineers, Vol. 60, No. 7, July 1990.

SOUTHWOOD, J. M., R. C. HARRIS, and O. MACKAY, "Modeling the Fate of Chemicals in an Aquatic Environment: The Use of Computer Spreadsheet and Graphics Software," *Environmental Toxicology and Chemistry,* Vol. 8, No. 11, 1989.

STARR, L. E., and K. E. ANDERSON, "A USGS Perspective on GIS," *Geographical Information Systems: Principles and Applications* (edited by Wright), Longman Scientific and Technical, 1991.

STILES, J. M., and R. E. PUNNETT, "Including Expert System Decisions in a Numerical Model of a Multi-Lake System Using STELLA," *Water Management in the '90s, a Time for Innovation, Proceedings of 20th Annual WRPMD Conference,* American Society of Civil Engineers, New York, 1993.

STOVER, E. L., and C. K. CAMPANA, "Computerized Biological Treatment Operational Process Control," *Water Science and Technology,* Vol. 24, No. 6, 1991.

TSAI, A. Y. H., *Database Systems, Management and Use,* Prentice Hall, 1988.

Universities Council on Water Resources, "Geographic Information System Issues," *Water Resources Update,* Issue 87, Winter 1992.

U.S. Army Corps of Engineers, "Object Oriented Simulation Modelling for Water Resources," Draft Engineer Technical Letter, Washington, D.C., April 1991.

U.S. Army Corps of Engineers, Hydrologic Engineering Center, "HECDSS User's Guide and Utility Program Manuals," CPD-45, Davis, California, December 1990.

WALKER, J. F., S. A. PICKARD, and W. C. SONZOGNI, "Spreadsheet Watershed Modeling for Nonpoint-Source Pollution Management in a Wisconsin Basin," *Water Resources Bulletin,* American Water Resources Association, Vol. 25, No. 1, February 1989.

WERICK, W. J., "National Study of Water Management During Drought: Results Oriented Water Resources Planning," *Water Management in the '90s, a Time for Innovation, Proceedings of 20th Annual WRPMD Conference,* American Society of Civil Engineers, New York, 1993.

WRIGHT, J. R., L. L. WIGGINS, R. K. JAIN, and T. J. KIM (editors), *Expert Systems in Environmental Planning,* Springer-Verlag, 1993.

YATES, S. R., and M. V. YATES, "Geostatistics for Waste Management: A User's Manual for the GEOPACK (Version 1.0) Geostatistical Software System," EPA/600/8-90/004, U.S. Environmental Protection Agency, Robert S. Kerr Environmental Research Laboratory, Ada, Oklahoma, 1990.

YEH, W. W-G., "Reservoir Management and Operations Models: A State-of-the-Art Review," *Water Resources Research,* AGU, Vol. 21, No. 21, December 1985.

YEH,W. W-G., "Systems Analysis in Ground-Water Planning and Management," *Journal of Water Resources Planning and Management,* American Society of Civil Engineers, Vol. 118, No. 3, May/June 1992.

4

DEMAND FORECASTING AND BALANCING SUPPLY WITH DEMAND

INTRODUCTION

Each of Chapters 4 through 10 addresses a particular type or category of water management models and modeling applications. Various aspects of managing water supplies are addressed in several of the chapters. This chapter addresses the topics of water use forecasting and balancing water demand and supply. It focuses on two models: IWR-MAIN and WEAP. IWR-MAIN is a municipal and industrial water use forecasting system, which includes capabilities for estimating the effectiveness of demand management practices. The Water Evaluation and Planning (WEAP) System is a water demand and supply accounting model, which provides capabilities for comparing water supplies and demands as well as for forecasting demands. IWR-MAIN and WEAP are included in the Model Inventory Appendix.

MUNICIPAL AND INDUSTRIAL WATER USE FORECASTING

Water resources planning and management is highly dependent on projections of future water needs. Urban water use projections are required in the following:

- Planning and implementing expansions to existing facilities for supplying, treating, and distributing water and for collecting and treating wastewater

- Planning and implementing major new construction projects such as dams and conveyance facilities
- Planning and implementing water rights reallocations or reallocations of storage capacity in existing reservoirs
- Preparing drought contingency plans
- Evaluating the effectiveness of alternative demand management (water conservation) plans
- Implementing demand management measures during drought conditions
- Predicting utility revenues to be expected from the sale of water
- Developing local, regional, and national water resources assessments and formulating water management policies and plans

All of these activities are based on estimates of future water requirements. The future may be measured in days, years, or decades. Major constructed facilities are planned over many years, or even decades, to meet needs extending many decades into the future. On the other hand, implementation of demand management measures during drought conditions may focus on water needs during the next several days or weeks.

Water use in an urban area is dependent on various demographic, climatic, and socioeconomic factors such as

- Resident and seasonal population
- Personal income
- Climate
- Weather conditions
- Number, market value, and types of housing units
- Employment in service industries
- Manufacturing employment and output
- Water and wastewater prices and rate structures
- Irrigated acreage in residential, commercial, and public use
- Types of lawns and watering practices
- Water-using appliances
- Demand management activities

Municipal and industrial water use forecasting is based on relating water use to estimates of future values of one (such as population) or more (possibly several or all of the foregoing) water use determinants.

Demand management or water conservation programs represent a key determinant of water use. Demand management has also provided a major impetus for improving and refining water use forecasting methods in recent years. Prior to the late 1970s, water supply planning and management was based essentially on increasing dependable supplies as necessary to meet projected demands. A major water policy thrust of the late 1970s and 1980s was to shift to a greater reliance on reducing demands by improving use efficiency instead of relying solely on augmenting supplies. In recent years, methods for forecasting water use and for evaluating water conservation plans have been closely interrelated. Water use forecasting methods now typically include capabilities for reflecting alternative demand management strategies in forecasts.

Water Use Forecasting Literature

During the late 1970s and early 1980s, the Institute for Water Resources of the U.S. Army Corps of Engineers conducted a research and information transfer program to develop methods for evaluating municipal and industrial water conservation plans, including improving water use forecasting capabilities. The development of methods for formulating and evaluating water conservation and drought management measures are documented by Baumann et al. (1979a, 1979b, 1980, 1981), Boland et al. (1981a, 1981b), Crews and Tong (1981), and Dziegielewski et al. (1983, 1993). Dziegielewski et al. (1981) and Boland et al. (1981b) developed an annotated bibliography and an assessment of water use forecasting methods and provided a comprehensive literature review. The IWR-MAIN model stemmed from this series of studies (Crews and Miller 1983; Davis et al. 1991).

Mays and Tung (1992) provide a textbook overview of municipal and industrial water use forecasting, including a brief description of IWR-MAIN and a review of regression analysis and other related statistical methods. Goodman (1984) reviews methods for estimating population and water needs in general. Shaw and Maidment (1987, 1988) discuss a microcomputer model for short-term forecasting of water use and evaluating the effectiveness of water conservation programs.

Although water use is addressed extensively in the research literature, most studies involve somewhat ad hoc projections for a particular location. Only a limited number of comprehensive, in-depth research efforts have been reported in regard to developing basic water use data and relationships between water use and explanatory determinants of water use.

A classic study of residential water use was conducted at Johns Hopkins University from 1961 to 1966 to determine the water use patterns and demand rates imposed on water systems in residential areas and to define the major factors influencing residential water use (Howe and Linaweaver 1967). A number of other subsequent studies have used information from this early research project. Most of the residential water use submodels contained in the IWR-MAIN software package are based on data from the Johns Hopkins University study.

The residential water use research project conducted at Johns Hopkins University was sponsored by the Federal Housing Administration in cooperation with 16 participating utilities from throughout the United States. Master-meter, punched-tape recorder systems were installed to monitor continuously water flow into 39 homogeneous residential areas served by the 16 water utilities. The 39 study areas ranged in size from 34 dwelling units to 2,373 dwelling units. Howe and Linaweaver (1967) used the data collected to analyze average daily use, maximum daily use, peak hourly use, and indoor versus outdoor use. A comprehensive econometric analysis resulted in water use models for five categories of residential water use:

- Metered and sewered residences in the western United States
- Metered and sewered residences in the eastern United States
- Metered residences with septic tanks
- Flat rate and sewered residences
- Apartment areas

For each housing category, regression equations were developed for domestic demand, summer sprinkling demand, and maximum-day demand. Some of the water use determinants incorporated in the regression equations included

- Market value of residence
- Number of persons per dwelling unit
- Age of dwelling unit
- Average water pressure
- Marginal water and sewer charge
- Irrigable area per dwelling unit
- Summer potential evaporation in inches

In 1982, Howe reestimated some of the original Howe and Linaweaver (1967) equations to incorporate a new water use determinant called the bill difference variable. The bill difference is calculated as the difference between the total water/wastewater price and water use. The revised equations are incorporated in IWR-MAIN.

Water Use Forecasting Methods

Water use forecasting can be characterized by (1) the level of complexity of the mathematical relationships between water use and explanatory variables or determinants of water use, and (2) the level of sectoral, spatial, seasonal, and other disaggregation of water users. The complexity of the relationships depends primarily on how many and which explanatory variables are included in the equations. Disaggregation refers to making separate estimates for categories and subcategories of water use. For example, sectoral disaggregation involves separate water use predictions for residential, commercial, industrial, institutional, and public uses, which, in turn, can each be divided into numerous subcategories. The separate water use forecasts are aggregated or added together to obtain the total water use.

One of the simplest and typically least accurate methods for forecasting future water use is to extrapolate historical water use data. The historical change in water use is extrapolated into the future, using graphical or mathematical means, without considering expected trends in the factors that actually determine water use.

Another conceptually simple, routinely used approach is to estimate water use (Q_t) at a future time (t) by multiplying the future population (P_t) by a per capita water use rate (r_t) as follows:

$$Q_t = r_t \times P_t$$

The per capita water use rate can be assumed constant or projected to change over time based on historical water use data. Per capita water use rates can be estimated based on water use records for a particular city; or, alternatively, regional or national use rates can be obtained from the literature.

Per customer or per connection methods are a variation of the per capita approach. Future water use is the product of the projected number of customers and a projected value of water use per customer. This approach is used most frequently in conjunction with sectorally disaggregate forecasts, where water use per customer coefficients are estimated for each customer class. Thus, water use forecasts can reflect varied growth rates among customers.

Commercial and industrial water use is commonly forecast on a per employee basis. Disaggregated forecasts for specific sectoral categories are frequently expressed as a single coefficient function of other variables, such as number of hotel rooms or hospital beds.

Water use forecasting models are often based on regression equations which relate mean or peak water use rates to one or more determinants of water use (explanatory variables). A typical general form of the regression equations is as follows:

$$Q = a + bX_1^e + cX_2^f + \ldots + dX_n^g$$

where

$$Q = \text{forecasted water use rate}$$

$$X = \text{explanatory variables}$$

n = number of explanatory variables

a,g = regression coefficients or parameters

Typical examples of explanatory variables include resident and seasonal population; personal income; number, market value, and types of housing units; employment; manufacturing output; water and wastewater prices and rate structures; irrigated acreage; climate (arid or humid); weather conditions; and water conservation programs. A disaggregated forecast may involve any number of equations representing various categories and subcategories of water use. In general, greater forecast accuracy and greater flexibility in representing alternative future scenarios and management strategies can be achieved by increased disaggregation and inclusion of more explanatory variables in the forecast equations. The exact form of each forecast equation and value for the coefficients can be determined from regression analyses of past water use data for the particular study area. Alternatively, generic equations have been developed based on data from many study areas representative of geographic regions or the entire nation.

Water use forecasting methods are sometimes differentiated as being either requirements models or demand models. Requirements models do not include the price of water or other economic factors as explanatory variables, thus implying that water use is an absolute requirement unaffected by economic choice. Demand models include the price of water to the user as an explanatory variable, as well as related economic variables such as income.

The aforementioned water use forecasting methods are based on projections of future values for the determinants of water use. Data are also required to develop the coefficients in the regression equations. Thus, data availability is a key consideration in water use forecasting. Data are available from a variety of sources. For example, historical data and future projections related to population, personal income, housing, and employment can be obtained from published census data and OBERS regional projections; local and state planning agencies; econometric firms; and state and national statistical abstracts. Climate data are available from National Weather Service publications as well as from various federal, state, and local agencies. Water use data for the study area and information regarding local water and wastewater pricing and water conservation programs are obtained from water utilities and local agencies.

IWR-MAIN Water Use Forecasting System

The IWR-MAIN Water Use Forecasting System is a software package which provides a variety of forecasting models, socioeconomic parameter generating procedures, and data management capabilities. The acronym IWR-MAIN stands for Institute for Water Resources—Municipal and Industrial Needs.

IWR-MAIN was originally based on the MAIN model developed by Hittman Associates, Inc. in the late 1960s for the U.S. Office of Water Resources Research, which was in turn based on earlier work by Howe and Linaweaver (1967) and others. In the early 1980s, the IWR adopted and modified MAIN and renamed the revised model IWR-MAIN. During the 1980s, IWR-MAIN evolved through several versions representing major modifications. Version 5.1, documented by Davis et al. (1991), has recently been replaced by Version 6.0. The model is available by contacting IWR or Planning and Management Consultants, Ltd. (PMCL). PMCL periodically offers a training course on application of IWR-MAIN, in coordination with IWR and the American Public Works Association. Future plans include distribution of the software by the American Public Works Association through a users group. IWR-MAIN has been applied to a number of cities located throughout the United States.

IWR-MAIN is a flexible municipal and industrial water use forecasting system. Forecasts are made for average daily water use, winter daily water use, summer daily water use, and maximum-day summer water use. IWR-MAIN provides capabilities for highly disaggregated forecasts. Water requirements are estimated separately for the residential, commercial/institutional, industrial, and public/unaccounted sectors. Within these major sectors, water use estimates are further disaggregated into categories such as metered and sewered residences, commercial establishments, and three-digit SIC manufacturing categories. A maximum of 284 categories can be accommodated, but most forecasts utilize approximately 130 specific categories of water use.

Water use is estimated as a function of one or more explanatory variables, which may include

- Number of users
- Number of employees in nonresidential categories
- Price of water and sewer service
- Market value of housing units in residential categories
- Number of persons per housing unit in residential categories
- Climate and weather conditions
- Conservation programs

Different subsets of explanatory variables are used to estimate average, average winter, average summer, and maximum-day summer water use in various water use sectors and categories. A mixture of water use estimating equations is employed in the computational routines of IWR-MAIN. For some user sectors, water use is predicted with econometric demand models, including price in the case of residential users with water meters. Other uses are estimated by means of requirements models, usually of the unit use coefficient type.

IWR-MAIN contains a procedure for estimating the water-saving effectiveness of water conservation (demand management) programs. Conservation parameters obtained from literature sources are provided for 14 measures:

- Public information/education program
- Metering of customer connections
- Reduction of system pressure
- Water rate policy changes
- Rationing program
- Sprinkling restrictions
- Industrial reuse and/or recycling
- Commercial reuse and/or recycling
- Leak detection and repair
- Retrofit of showers and toilets
- Moderate plumbing code
- Advanced plumbing code
- Low-water-use landscaping for new construction
- Retrofit low-water-use landscaping

The impacts of one or more proposed or previously implemented conservation measures in the water service area are computed based on (1) estimates of the expected reduction in the uses of water affected by conservation, (2) the market coverage of conservation practices, and (3) expected interactions among measures that are implemented together.

Preparation of an IWR-MAIN water use forecast requires (1) verification of the empirical equations and coefficients for estimating water use and (2) projection of future values of determinants of water use. Model verification is accomplished by preparing independent estimates of water use for one or more historical years and comparing these estimates with actual water use conditions. If necessary, the model can be calibrated.

The base year is the year from which values of explanatory variables are projected. A calendar year that coincides with the U.S. Census of Population and Housing is typically selected as the base year. One or more subsequent years are selected as the forecast years for which water use is predicted.

Future values of water use determinants can be developed externally or can be generated by growth equations built into the program. However, not all future parameters can be generated by the internal growth models. Total population, total employment, and median household income in each forecast year must be provided by the user. The growth models for the residential sector can produce default projections of total number of housing units and their distribution by market value from (1) base year housing data, (2) the projected median household income, and (3) population growth rate. Similarly, the distribution of employment among eight major SIC divisions is projected for each forecast year, using base year values and past employment trends in each category.

The recent Version 6.0 of IWR-MAIN includes a module called the integrated water supply and demand plan. Capabilities are provided for selecting a least-cost combination of water supply and demand management alternatives in response to deficits between baseline forecasts of water use and expected yields of supply sources. Trade-offs can be evaluated between the investment in long-term demand and supply management alternatives and the costs of coping with periodic shortages of supply during drought conditions.

AGRICULTURAL WATER REQUIREMENTS

Forecasting agricultural water use involves predicting future cropping patterns and management practices and estimating evapotranspiration and water needs for particular crops. The American Society of Agricultural Engineers (1990) addresses a comprehensive range of irrigation management topics, including modeling of evapotranspiration, crop growth, agricultural economics, and irrigation scheduling. Burman et al. (1983) and the American Society of Civil Engineers (1990) outline methods for estimating irrigation water requirements. The U.S. Bureau of Reclamation (1991) describes 14 computer models categorized as water requirements models. These models deal with estimating evapotranspiration and crop water requirements and managing irrigation. Most of the models were developed by the Bureau of Reclamation. Some are site specific, but most are generalized for application to various locations.

WATER EVALUATION AND PLANNING SYSTEM

The Water Evaluation and Planning (WEAP) model is a water demand and supply modeling system which serves several purposes, including database management, forecasting, and analysis. WEAP provides a database system for maintaining water demand and supply information. It provides capabilities for forecasting water demand and supply over a long-term planning horizon. It is a simulation model for evaluating alternative water use scenarios and management strategies. WEAP can be used to perform various types of analyses, including sectoral water demand forecasts, supply source allocations, streamflow and reservoir storage simulations, hydropower forecasts, pollution loading estimates, and benefit-cost analyses. WEAP can be applied to single or multiple interconnected river systems at the city, regional, or national level.

WEAP was developed by the Tellus Institute, which is a team of scientists, planners, and policy analysts organized into a nonprofit research and consulting organization. The Tellus Institute serves as the Boston Center of the Stockholm Environment Institute, an international organization based in Sweden. The 1993 version of the WEAP model (Tellus Institute 1993) expands the original 1990 version and continues to be refined. WEAP has been applied in studies in several countries. A study by the Hydrologic Engineering Center (1993) is one of the first applications in the United States. This case study was performed in conjunction with a HEC research effort to test, enhance, and apply the WEAP modeling package.

WEAP runs on MS-DOS-based microcomputers in an interactive, menu-driven mode. The model includes the following modules or programs: setup, demand, distribution, supply, and evaluation.

The setup program characterizes the problem under study by defining the study time period, physical elements comprising the water demand-supply network, and their spatial relationships. WEAP uses a network of nodes and arcs to represent the configuration of a water demand-supply system. A node represents a physical component such as a demand site (a water user or group of users) or a reservoir or other supply source. Nodes are linked by arcs, which represent natural or human-made flow connections such as river channels, canals, and pipelines. WEAP accounts for flow through each arc and at each node during each time interval of a simulation.

The purpose of the demand program is to forecast water demands for various water uses defined in the study. Projected water demands determined in this program are passed to the distribution, supply, and evaluation programs for further processing and analysis. The model user inputs information regarding present and future socioeconomic development measures (such as population, industrial output, agricultural output, and urban and rural domestic development) and unit water use requirements (per capita, per production output, or per activity in general). The program computes the water demands over time by multiplying every activity measure with its unit water requirement. Both the activities and unit water requirements may vary with time. Three optional methods are provided for projecting activity levels and water use rates: interpolation, drivers and elasticities, and growth rate. Drivers are the explanatory variables chosen for the water use projections, such as population, consumption, industrial output, or investment. Elasticities are used in conjunction with drivers to model activity levels or water use rates that do not change proportionally to the drivers. Elasticities are defined by econometric relationships.

The demand program uses a hierarchical branching structure to manage data. The levels are sector, subsector, end-use, and device. The economy is divided into sectors, which are defined for the particular study. Sectors for a typical study might include agriculture, industry, and municipal. Each sector is divided into subsectors. For example, the industrial sector could be disaggregated by various industrial classifications such as petrochemical, textile, electric power, etc. The agricultural sector might be disaggregated by type of crop. Each subsector is divided into end-uses. For example, a crop subsector might be characterized by water requirements in different soil conditions or in different locations in the study area. Each end-use is further divided into the devices used, which could include irrigation techniques, toilet models, or cooling processes. Each device is linked to one or more demand sites.

The distribution program converts the annual demands developed in the demand program into monthly supply requirements by incorporating monthly variation coefficients, distribution losses, conveyance capacities, and reuse rates for each demand site. Losses could reflect pipeline leaks, canal seepage and evaporation, clandestine connections, or unmetered water. Demand site losses are specified as a percentage of demand. Return flows are specified as a percentage of withdrawals and may be returned to either rivers or ground water.

The supply program simulates the spatial and temporal water allocations between supply sources and demand sites. Two broad types of supply sources are reflected in the model: local sources and river sources. Local sources represent all sources where a streamflow account is not needed, such as a ground-water source, interbasin transfer, or diversion from a reservoir not dependent on river flow. The capacities of ground water and other local sources are represented by inputted firm yields. When the river simulation mode is active, streamflows are tracked along a main river and its tributaries. Two optional methods are provided for specifying streamflows: historical data method and simplified method. Historical streamflows for each month of the simulation period can be inputted for all pertinent locations. Alternatively, the simplified method allows use of data for five types of years: very wet, wet, normal, dry, and very dry. Reservoir storage is divided into four zones: inactive, buffer, conservation, and flood control.

The evaluation program provides capabilities for comparing and evaluating alternative water use scenarios and management strategies in terms of physical demand and supply, environmental impacts, and economic costs and benefits. The program organizes and displays inputted and computed data comparing alternative plans.

REFERENCES

American Society of Agricultural Engineers, *Management of Farm Irrigation Systems,* ASAE Monograph, edited by G. J. Hoffman, T. A. Howell, and K. H. Solomon, St. Joseph, Missouri, December 1990.

American Society of Civil Engineers, *Evaporation and Irrigation Water Requirements,* Manual No. 70, edited by M. E. Jensen, New York, 1990.

BAUMANN, D. D., J. J. BOLAND, J. H. SIMS, B. KRANZER, and P. H. CARVER, "The Role of Conservation in Water Supply Planning," Contract Report 79-2, USACE Institute for Water Resources, Fort Belvoir, Virginia, April 1979a.

BAUMANN, D. D., K. ALLEY, J. BOLAND, P. CARVER, B. KRANZER, and J. SIMS, "An Annotated Bibliography on Water Conservation," Contract Report 79-3, USACE Institute for Water Resources, Fort Belvoir, Virginia, April 1979b.

BAUMANN, D. D., J. J. BOLAND, and J. H. SIMS, "The Evaluation of Water Conservation for Municipal and Industrial Water Supply—Procedures Manual," Contract Report 80-1, USACE Institute for Water Resources, Fort Belvoir, Virginia, April 1980.

BAUMANN, D. D., J. J. BOLAND, and J. H. SIMS, "The Evaluation of Water Conservation for Municipal and Industrial Water Supply—Illustrative Examples," Contract Report 82-C1, USACE Institute for Water Resources, Fort Belvoir, Virginia, February 1981.

BOLAND, J. J., B. DZIEGIELEWSKI, D. D. BAUMANN, and E. M. OPITZ, "Influence of Price and Rate Structures on Municipal and Industrial Water Use," Contract Report 84-C-2, USACE Institute for Water Resources, Fort Belvoir, Virginia, May 1981a.

BOLAND, J. J., D. D. BAUMANN, and B. DZIEGIELEWSKI, "An Assessment of Municipal and Industrial Water Use Forecasting Approaches," Contract Report 81-C05, USACE Institute for Water Resources, Fort Belvoir, Virginia, May 1981b.

BURMAN, R. D., R. H. CUENCA, and A. WEISS, "Techniques for Estimating Irrigation Water Requirements," *Advances in Irrigation* (D. Hillel), Vol. 2, Academic Press, New York, 1983.

CREWS, J. E., and J. TONG (editors), "Selected Works in Water Supply, Water Conservation and Water Quality Planning," Research Report 81-R10, USACE Institute for Water Resources, Fort Belvoir, Virginia, May 1981.

CREWS, J. E., and M. A. MILLER, "Forecasting Municipal and Industrial Water Use, IWR MAIN System User's Guide for Interactive Processing and User's Manual," Research Report 83-R-3, USACE Institute for Water Resources, Fort Belvoir, Virginia, July 1983.

DAVIS, W. Y., D. M. RODRIGO, E. M. OPITZ, B. DZIEGIELEWSKI, D. D. BAUMANN, and J. J. BOLAND, "IWR-MAIN Water Use Forecasting System Version 5.1," Report 88-R-6, USACE Institute for Water Resources, Fort Belvoir, Virginia, June 1988, Revised August 1991.

DZIEGIELEWSKI, B., J. J. BOLAND, and D. D. BAUMANN, "An Annotated Bibliography of Techniques of Forecasting Demand for Water," Contract Report 81-C03, USACE Institute for Water Resources, Fort Belvoir, Virginia, May 1981.

DZIEGIELEWSKI, B., D. D. BAUMANN, and J. J. BOLAND, "The Evaluation of Drought Management Measures for Municipal and Industrial Water Supply," Contract Report 83-C-3, USACE Institute for Water Resources, Fort Belvoir, Virginia, December 1983.

DZIEGIELEWSKI, B., E. M. OPITZ, J. C. KIEFER, and D. D. BAUMANN, "Evaluating Urban Water Conservation Programs: A Procedures Manual," American Water Works Association, Denver, Colorado, 1993.

GOODMAN, A. S., *Principles of Water Resources Planning,* Prentice Hall, 1984.

HOWE, C. W., and F. P. LINAWEAVER, "The Impact of Price on Residential Water Demand and Its Relation to System Design and Price Structure," *Water Resources Research,* Vol. 3, 1967.

Hydrologic Engineering Center, "Accounting for Water Supply and Demand: An Application of Computer Program WEAP to the Upper Chattahoochee River Basin, Georgia," Draft, TD-34, U.S. Army Corps of Engineers, Davis, California, September 1993.

MAYS, L. W., and Y-K. TUNG, *Hydrosystems Engineering & Management,* McGraw-Hill, 1992.

SHAW, D. T., and D. R. MAIDMENT, "Forecasting Water Use in Texas Cities," TR-125, Texas Water Resources Institute, College Station, Texas, August 1987.

SHAW, D. T., and D. R. MAIDMENT, "Effects of Conservation on Daily Water Use," *Journal of the American Water Works Association,* Vol. 80, September 1988.

Tellus Institute, Stockholm Environment Institute Boston Center, "WEAP, a Computerized Water Evaluation and Planning System, User Guide," Boston, Massachusetts, April 1993.

U.S. Bureau of Reclamation, "Inventory of Hydrologic Models," Global Climate Change Response Program, Denver, Colorado, August 1991.

5

WATER DISTRIBUTION
SYSTEM MODELS

INTRODUCTION

This chapter addresses the hydraulic engineering problem of modeling flows and pressures in pipe networks. Many computer programs have been developed to perform the fundamental pipe network analysis computations. This chapter focuses on KYPIPE2, developed at the University of Kentucky, and WADISO (Water Distribution Simulation and Optimization), developed at the USACE Waterways Experiment Station, which are widely applied, user-oriented comprehensive water distribution system analysis models.

Model Applications

Analysis of municipal water distribution systems represents a major modeling application. Other types of pipe networks frequently modeled include industrial water conveyance systems, rural water supply systems, sprinkler systems, and surcharged storm sewer systems. Distribution and conveyance systems include pipes, pumps, storage tanks, valves, and various pipe fittings. Although this discussion focuses on water, the modeling techniques are also applicable for other liquids, such as petroleum and chemical products. Models are applied in the investigation of existing facilities, proposed extensions and modifications to existing systems, and, in some cases, proposed new pipe networks. Urban growth results in continuing expansions of municipal systems in many areas. Rehabilitation of aging water distribution systems is a major infrastructure concern throughout the United States.

Models simulate the impacts of various water demand scenarios on pressures and flows throughout a system. For example, the impact of a new residential development on system capabilities for meeting demands and maintaining pressures throughout the water utility service area may be of concern. Simulations are likewise performed to analyze the impacts of alternative system improvements, such as new pipes, pumps, or storage tanks. Models are used to size pipes, select pumps, and otherwise design systems. Modeling studies have been performed to develop pump operating strategies which meet water demands while minimizing electrical energy costs associated with pump operation. Calibration or estimation of values for head loss coefficients and other model parameters represents another type of model application.

Hydraulic Analysis of Pipe Networks

A simulation model typically computes discharges in each pipe and the hydraulic grade (elevation plus pressure head) and associated pressure at each node of a pipe network. Nodes represent pipe junctions, water use demand locations, and/or locations of known or computed hydraulic heads. Known hydraulic heads are inputted for fixed grade nodes, which may represent storage tanks or water mains supplying the system at a known head. The pipe network must have at least one known or fixed grade node in order for the model to compute hydraulic grades at all the other nodes. Input data include length, diameter, and roughness coefficient for each pipe; minor loss coefficients for valves and other fittings; elevations of the junction nodes; head versus discharge relationships characterizing each pump; and external water demands at various nodes in the system.

Pipe network analysis computations are based on the following concepts:

- The flow entering each junction must equal the flow leaving it in accordance with the continuity equation (conservation of mass).
- The algebraic sum of the head losses and gains (pumps) around any closed pipe loop must be zero in accordance with conservation of energy.
- The algebraic sum of the head losses and gains (pumps) along any path between two fixed grade nodes must equal the difference in head between the two fixed grade nodes.

Head losses in pipes are estimated using either the Hazen-Williams or Darcy-Weisbach equations. Additional "minor" losses due to valves and other appurtenances are estimated by applying an inputted empirical loss coefficient to the velocity head.

A set of equations reflecting the aforementioned fundamental concepts is formulated and solved to determine the flow rate in each pipe or the hydraulic head at each node. After computing the discharges in each pipe, the hydraulic heads at the junctions are determined, or vice versa. Some models, such as KYPIPE2, use the loop method, in which flows in each of the pipes are computed first. Then the hydraulic grade is computed at each node given the computed flows and at least one inputted head. Other models, such as WADISO, use the node approach of first computing the hydraulic grades at each node and then the flows in each pipe. Because head loss terms in the equations are nonlinear, iterative solution algorithms are required with any approach. All pipe network analysis models are based on the fundamental concepts of conservation of mass and energy, as discussed earlier. Alternative models differ in the node versus loop approaches for formulating the solution algorithms and the iterative methods used to solve the set of nonlinear equations. Early models are based on the Hardy Cross method of iteratively performing computations for each individual loop in turn. Most more recent models, including KYPIPE2 and WADISO, use linearization schemes which approximate and iteratively adjust nonlinear terms in the governing equations. The resulting system of linear equations is solved simultaneously for each iteration.

Models may also be formulated to compute pipe diameters, roughness coefficients, or other system parameters, given inputted hydraulic head and water demand requirements. Pipes may be sized or pump operation strategies designed based on minimizing cost. WADISO and KYPIPE2 use enumeration approaches based on iterative executions of the simulation model for alternative pipe sizes or other decision variables. Formal optimization (mathematical programming) algorithms have been used in other models.

WADISO and KYPIPE2 provide capabilities for either steady-state or extended period simulations. Steady-state models describe the hydraulics of a system at an instant in time, without consideration of changes over time. In an extended period simulation, filling and drainage of storage tanks are modeled. An extended period simulation involves repeating steady-state simulations at discrete intervals of time represented by different water levels in storage tanks and perhaps different water demands. For a given tank water level and inflow and outflow rates at a point in time, the tank water level at a later time can be estimated reasonably accurately as long as the time interval is relatively short and changes in storage and other variables are relatively slow.

When velocities and momentum in pipes change quickly due to rapid valve closure or other causes, pressure surges travel through the pipe system. These rapid changes in momentum, associated with water hammer and other problems, are modeled with a different set of equations and computational techniques than the steady-state and extended period simulations covered in this chapter. Watters (1984) and Wylie and Streeter (1993) treat unsteady or transient flow in pipe systems and provide a set of computer programs for applying various analysis techniques.

REVIEW OF AVAILABLE MODELS

Jeppson (1976) and Walski (1984) provide comprehensive treatments of water distribution network analysis methods. Wood and Rayes (1981) compare alternative algorithms for solving the basic hydraulic equations. Male and Walski (1990) provide a handbook for solving practical problems encountered in the operation and maintenance of water distribution systems. The Task Committee on Risk and Reliability Analysis of Water Distribution Systems (1989) covers a variety of analysis strategies and methods, including both simulation and optimization models. Goulter (1992) reviews the application of optimization techniques, such as linear and nonlinear programming, in the design and analysis of water distribution networks and concludes that, although pipe system simulation models are routinely applied by practitioners, formal optimization methods have been used primarily in academic research studies.

Most water distribution system simulation models are conceptually similar. Many models, incorporating the basic concepts discussed earlier, have been developed and applied in specific studies without being adopted by users other than the original model developers. Other operational generalized models have been widely applied. KYPIPE2, discussed later, is probably the most widely used of the available models. CYBERNET, available from Haestad Methods (Waterbury, Connecticut, 800/727-6555), is an AutoCAD-based version of KYPIPE2. CYBERNET has enhanced graphical user interface features. WADISO, also discussed later, is a comprehensive, user-friendly modeling package which is similar to KYPIPE2. Other generalized programs with similar capabilities include the WATER and WATEXT models developed by the U.S. Bureau of Reclamation (1991) and the UNWB-LOOP model developed by the World Bank (1985). Wunderlich and Giles (1986) compare three pipe network models (KYPIPE, WADISO, and UNWB-LOOP) and also two general-purpose problem solvers (TK!SOLVER and MINOS) applied to pipe networks. Helweg (1991) provides computer codes written in BASIC for (1) analyzing flow in pipe networks using the Hardy Cross method and (2) computing head losses in a single pipeline.

MODELS INCLUDED IN THE MODEL INVENTORY
APPENDIX

WADISO and KYPIPE2 are included in the Model Inventory Appendix. The two computer programs provide similar modeling capabilities. Both are menu-driven, user-interactive modeling systems with executable versions available for MS-DOS-based microcomputers. KYPIPE2 is a proprietary computer program. WADISO is in the public domain.

KYPIPE2

The original KYPIPE (Wood 1980) and revised KYPIPE2 (Wood 1991) were developed at the University of Kentucky. The model has been widely applied throughout the United States and abroad by engineers working for cities, consulting firms, agencies, and universities. Short courses for practicing engineers on the use of the model are periodically offered by various universities and firms.

KYPIPE2 is a comprehensive system for modeling flow in pipe networks. The interactive, menu-driven software package includes the KYPIPE2 simulation program along with a program for preparing and editing input data and several programs for graphical displays of modeling results. The KYPIPE2 package also includes features to facilitate use of the AutoCAD computer-aided drafting system to prepare drawings of pipe networks. KYPIPE2 calculates steady-state flows and pressures for pipe distribution networks. Optional capabilities are also provided for extended period simulations with storage tank levels varying over time. The model will calculate flows for each pipe and the pressure at each node for a given set of water demands. Alternatively, capabilities are provided to compute, for specified pressure requirements, a variety of design, operation, and calibration parameters. The model will determine pump speed, pump power, hydraulic grade settings for storage tanks or supplies, hydraulic grade settings for regulating valves, control valve settings or loss coefficients, pipe diameters, roughness coefficients, and water demands or flow requirements.

KYPIPE2 is based on iteratively solving the full set of mass continuity and energy equations utilizing linearization schemes to handle nonlinear terms and a sparse matrix routine for solving the resulting set of linear equations. Pipe head losses are estimated using either the Hazen-Williams or Darcy-Weisbach equations.

WADISO

The Water Distribution Simulation and Optimization (WADISO) model was originally developed in conjunction with the Water Supply and Conservation Research Program of the U.S. Army Corps of Engineers Waterways Experiment Station (Gessler and Walski 1985; Walski and Gessler 1988). WADISO was applied to a number of systems worldwide by the Corps of Engineers. The model was later documented and distributed as a published book (Walski et al. 1990), which is accompanied by diskettes containing both the source code and executable program compiled for MS-DOS-based microcomputers. The book has three parts: (1) "Applying Water Distribution Models to Real Systems," (2) "Computer Analysis of Pipe Networks," and (3) "User's Guide to WADISO." Appendixes provide detailed instructions for running the microcomputer program.

WADISO consists of three modules: (1) simulation, (2) optimization, and (3) extended period simulation. The simulation module computes pressures at each node and flows in each pipe of a distribution system for specified water demands. The distribution system may contain pumps, pressure reducing valves, and check valves. Water demands may be specified at any of the nodes. There is essentially no limitation on the configuration of the pipe network. The Hazen-Williams equation is used to determine head losses in the pipes. The optimization module determines pipe sizes which will minimize cost while maintaining specified pressures and demands. The model uses a discrete enumeration algorithm to find the optimal pipe size combination within a user-specified range of parameters. The user groups the pipes to be sized. All pipes in a specified group are assigned the same diameter. The user also provides a list of discrete diameters to be considered and associated costs. The extended period simulation module computes flow and pressure distributions in a pipe network at discrete time intervals as tank water levels fluctuate and water use patterns vary.

REFERENCES

GESSLER, J., and T. M. WALSKI, "Water Distribution System Optimization," Technical Report EL-85-11, Environmental Laboratory, U.S. Army Engineer Waterways Experiment Station, Vicksburg, Mississippi, 1985.

GOULTER, I. C., "Systems Analysis in Water-Distribution Design: From Theory to Practice," *Journal of Water Resources Planning and Management,* American Society of Civil Engineers, Vol. 118, No. 3, May/June 1992.

HELWEG, O. J., *Microcomputer Applications in Water Resources,* Prentice Hall, 1991.

JEPPSON, R. W., *Analysis of Flow in Pipe Networks,* Ann Arbor Science Publishers, 1976.

MALE, J. W., and T. M. WALSKI, *Water Distribution Systems: A Troubleshooting Manual,* Lewis Publishers, 1990.

Task Committee on Risk and Reliability Analysis of Water Distribution Systems, *Reliability Analysis of Water Distribution Systems* (L. W. Mays, editor), American Society of Civil Engineers, New York, 1989.

U.S. Bureau of Reclamation, "Inventory of Hydrologic Models," Global Climate Change Response Program, Denver, Colorado, August 1991.

WALSKI, T. M., *Analysis of Water Distribution Systems,* Van Nostrand Reinhold Company, 1984.

WALSKI, T. M., and J. GESSLER, "Selecting Optimal Pipe Sizes for Water Distribution Systems," *Journal of the American Water Works Association,* Vol. 80, February 1988.

WALSKI, T. M., J. GESSLER, and J. W. SJOSTROM, *Water Distribution Systems: Simulation and Sizing,* Lewis Publishers, 1990.

WATTERS, G. Z., *Analysis and Control of Unsteady Flow in Pipelines,* Butterworth, 1984.

WOOD, D. J., "User's Manual, Computer Analysis of Flow in Pipe Networks Including Extended Period Simulations," University of Kentucky, Lexington, Kentucky, 1980.

WOOD, D. J., "KYPIPE2 User's Manual, Comprehensive Computer Modeling of Pipe Distribution Systems," Civil Engineering Software Center, University of Kentucky, Lexington, Kentucky, November 1991.

WOOD, D. J., and A. G. RAYES, "Reliability of Algorithms for Pipe Network Analysis," *Journal of the Hydraulics Division,* American Society of Civil Engineers, Vol. 107, No. HY10, October 1981.

World Bank, "Development and Implementation of Low-Cost Sanitation Investment Projects," UNDP Interregional Project INT/81/047, J.H. Read, Acting Project Manager, The World Bank, Washington, D.C., 1985.

WUNDERLICH, W. O., and J. E. GILES, "Review of Pipe Network Analysis," Report No. WR28-2-900-190, Tennessee Valley Authority, Norris, Tennessee, September 1986.

WYLIE, E. B., and V. L. STREETER, *Fluid Transients in Systems,* Prentice Hall, 1993.

6

GROUND-WATER MODELS

INTRODUCTION

Modeling ground-water systems involves both water quantity (flow) and water quality considerations. Ground-water models incorporate mathematical representations of some or all of the following processes: movement of water and other fluids through saturated or unsaturated porous media or fractured rock; transport of water-soluble constituents; transformation of contaminants by chemical, biological, and physical processes; and heat transport and associated effects of temperature variations on ground-water flow and pollutant transport and fate.

Model Applications

Ground-water modeling applications are typically motivated by water supply and/or water quality concerns. Models have also been used in studies of land subsidence due to ground-water pumping. Ground water and ground-water models may also play significant roles in managing environmental resources, such as ecological systems in rivers, estuaries, and wetlands. Although typically used to address specific water management concerns, models are also research tools used to develop a better generic understanding of ground-water systems and processes.

Ground-water flow models are often used in the planning, design, and management of well fields. Ground-water models are also applied in broader comprehensive planning studies of alternative water supply and demand management strategies. Models may be used to analyze water availability or water supply yields under various scenarios; drawdowns to be expected from alternative well construction and pumping plans; and effects of natural and human-induced recharge conditions. The impacts of saltwater encroachment or other constraints to water supply may be a motivating concern in modeling applications. Stream-aquifer interactions and conjunctive management of surface water and ground water may be a key concern in certain studies.

Regulatory activities for protecting ground-water quality provide a major impetus for developing and applying computer models. Contaminant transport models are used to evaluate the impacts of pollution from accidental spills, leaking storage tanks, underground wastewater injection, landfills, agricultural activities, mining operations, and various other sources. Ground-water models are used to evaluate designs for controlled waste management facilities. Models are also used in evaluating remediation plans for restoring contaminated ground-water systems. Ground-water models can also provide guidance in designing pollution monitoring systems required by federal, state, and local regulations.

Ground-water models may be functionally characterized as (1) descriptive prediction models, (2) more prescriptive management models, and (3) parameter identification models. Most ground-water models are prediction models which use inputted data characterizing the natural ground-water system and human-induced development to predict the hydraulic and/or water quality conditions to be expected for specified management scenarios. The commonly used descriptive ground-water models simulate flow, with or without solute transport, in continuous porous media. This general type of model is the focus of this chapter. Other more complex models simulate flow in fractured media, multiphase flow, or complex chemistry (geochemical models). The models provide descriptive predictions of the system response to a specified management plan. Prescriptive management models, on the other hand, are oriented toward determining optimal management plans for meeting specified objectives. These models involve linking an optimization algorithm with a flow and/or solute transport model. Parameter identification or calibration models address the inverse problem of determining values for the parameters which characterize the ground-water system. Optimization techniques are also incorporated in inverse models.

Modeling Fluid Flow and Solute Transport Processes

Ground-water modeling encompasses the interrelated processes of fluid flow, solute transport, and heat transport. Ground-water flow models simulate the movement of water and other fluids in porous or fractured rock systems. In addition to water, other fluids such as air or hydrocarbons are important in many modeling applications. Complex liquid pollutants may consist of multiple miscible and immiscible chemical components of varying density and viscosity. Saltwater intrusion is another density-driven flow phenomenon. The two dominant processes reflected in flow models are (1) flow in response to hydraulic potential gradients and (2) the loss or gain of water from sinks or sources, which include natural recharge or outflows, pumping or injection wells, and gains or losses in storage. Flow models have been developed for flow under saturated, unsaturated, and variable saturated/unsaturated conditions. Variably saturated models handle both conditions using a single set of equations. Other coupled saturated-unsaturated zone models have separate formulations for simulation of flow in the two zones.

Mathematical formulations for ground-water flow are based on the principles of conservation of mass and momentum. Conservation of mass is expressed by the continuity equation. Darcy's linear law for laminar flow provides an equation of motion which is applied in many saturated flow situations. Although analytical solutions are adopted for some relatively simple models, most complex ground-water models incorporate finite difference or finite element numerical solutions of the governing partial differential equations. In a numerical model, the continuous problem domain is replaced by a discretized domain consisting of an array of nodes and associated finite difference cells or finite elements. The nodal grid forms the framework of the numerical model. The hydraulic conductivity and storage parameters characterizing the ground-water system are assigned for each node, cell, or element. The governing equations are formulated for each grid location, and the distribution of heads and velocities is computed accordingly.

Boundary and initial conditions are specified in formulating a model for a particular ground-water system. Boundary conditions are mathematical statements specifying the dependent variable (head) or the derivative of the dependent variable (flux) at the boundaries of the problem domain. Physical boundaries of ground-water flow systems are formed by an impermeable body of rock or a large body of surface water. Ground-water divides form hydraulic or streamline boundaries. Boundary conditions for flow simulation may be categorized as (1) specified head, (2) specified flux, and (3) head-dependent flux. Transient (unsteady) flow models require specification of initial conditions, which consist of the distribution of head throughout the system at the beginning of the simulation. A steady-state head solution generated by a calibrated model is often used as the initial condition in transient models.

Contaminant transport models build on and expand fluid flow models. Transport of dissolved chemicals and biota such as bacteria and viruses is directly related to the flow of water. However, contaminant transport involves a significantly more complex array of processes. Many of the constituents occurring in ground water interact physically and chemically with each other and with the solid media. Solute transport involves both physical transport processes, such as advection and dispersion, and physical and chemical interactions, such as adsorption/desorption, ion exchange, dissolution/precipitation, reduction/oxidation, and radioactive decay. Biotransformations can also significantly alter the composition of ground water.

Solute transport models are based on applying the principle of mass conservation to each of the chemical constituents of interest. The resulting equations represent the physical, chemical, and biotic processes and interactions between the dissolved constituents and the solid surface matrix and among the various solutes themselves. In some cases, state equations are included in the model to reflect the influence of temperature and solute concentration variations of the fluid flow through the effect of these variations on density and viscosity.

In cases of high contaminant concentrations in wastewaters or highly saline water, changes in concentrations affect the flow patterns through changes in density and viscosity, which in turn affect the movement and spreading of the contaminant and hence the concentrations. In this situation, the model must incorporate either simultaneous or iterative solution of the flow and solute transport equations. In other cases involving low contaminant concentrations and negligible difference in specific weight between the contaminant and water, the ground-water model may be composed of a flow submodel and quality submodel. The flow model computes the piezometric heads. The water quality model then uses the head data to generate velocities for advective displacement of the contaminant, allowing for additional spreading through dispersion and for transformations by chemical and microbial reactions.

Heat transport is also considered in some ground-water models. Heat transport affects flow and contaminant transport processes. Conversely, heat transport may be significantly affected by other physical and chemical processes. The heat transport equation is derived by applying the energy balance principles concerning the transport, storage, and external sources/sinks of heat. Dependent variables may be temperature or enthalpy.

Few ground-water flow and contaminant transport problems can be modeled with confidence (National Research Council Committee on Ground Water Modeling Assessment 1990). The most satisfactory results to date have been achieved with models involving the flow of water or the transport of a single nonreactive contaminant in a saturated porous medium. The processes that control saturated ground-water flow are reasonably well understood, and standard models of these processes are generally believed to provide capabilities for obtaining reliable predictions if provided with adequate data. The heterogeneity of real-world ground-water systems is a major concern in developing the required input data. As systems become more complicated due to unsaturated conditions, fracturing, the presence of several mobile fluids, or the existence of reacting contaminants, many more questions arise about the adequacy or validity of the underlying process models.

Model Categorization

Ground-water flow models can be classified in various ways. Models can be classified based on spatial dimension as one-, two-, or three-dimensional, or quasi three-dimensional. Models can be either steady state or transient (time varying) or may optionally consider either situation. Models may be limited to fluid flow or consider solute transport and/or heat transport as well. Various physical, chemical, and biological transformation processes are included in different models. Models may consider saturated or unsaturated flow or both. Essentially all ground-water models deal with porous media, and a few have additional features for modeling flow through fractured rock.

Ground-Water Modeling Literature

Ground-water hydrology and hydraulics and contaminant hydrogeology are covered in books by Freeze and Cherry (1979), Domenico and Schwartz (1990), and Fetter (1993). Ground-water modeling principles and practices are outlined in detail by Bear and Verruijt (1987), Bear and Bachmat (1990), and Anderson and Woessner (1992). Anderson et al. (1993) provide a concise overview of subsurface water models and highlight several widely applied generalized software packages. Walton (1992) provides a guide for applying a variety of microcomputer utility programs in ground-water modeling. The National Research Council Committee on Ground Water Modeling Assessment (1990) provides a state-of-the-art assessment of ground-water modeling from the perspective of capabilities for supporting legal and regulatory activities mandated by various environmental legislation. Yeh (1992) reviews optimization methods which have been applied in ground-water planning and management particularly in regard to the inverse problem of parameter calibration. Javandel et al. (1984) review contaminant transport models. Istok (1989) outlines the finite element method as applied to ground-water modeling. Geraghty and Miller, Inc. (1991) maintains a bibliography of various categories of references on ground-water modeling. Van der Heijde et al. (1988) provide an overview of ground-water modeling and mention approximately 250 generalized models available from various sources. Van der Heijde and Elnawawy (1993) provide a comprehensive compilation of ground-water models.

REVIEW OF AVAILABLE MODELS

Generalized ground-water modeling software developed by various entities is available from a variety of sources. Computer programs available through various organizations and those in published books are reviewed.

Computer Programs in Published Books

Most complex ground-water models are documented by reports, user manuals, and papers devoted specifically to a particular model. However, several books have been published which provide and document relatively simple computer codes as well as discuss various aspects of ground-water modeling. The computer programs are furnished on diskettes accompanying the books or are available upon request. Clarke (1987) provides a collection of microcomputer programs and subroutines, coded in BASIC, which cover a variety of computations involved in modeling ground-water flow. Helweg (1991) provides several relatively simple programs, written in BASIC, for pump tests and pump selection. Bonn and Rounds (1990) published a program, called DREAM, which contains several routines for analytical solutions of relatively simple ground-water flow problems. Aral (1990) provides the Steady Layered Aquifer Model (SLAM) FORTRAN computer program. Walton (1989a) distributes four programs, coded in BASIC, called WELFUN, WELFLD, CONMIG, and GWGRAF, which provide analytical solutions to ground-water flow and contaminant migration problems. Walton (1989b) presents two numerical flow and contamination microcomputer programs, called GWFL3D and GWTR3D, which are modifications and extensions of the PLASM and RANDOM WALK models cited later in this chapter. Walton (1992) provides a tutorial and reference guide for applying MODFLOW, MODPATH, MODPATH-PLOT, MOC, SUTRA, INTERTRANS, INTERSAT, and GEOPACK.

The textbook by Bear and Verruijt (1987) is supplemented with a set of computer programs, coded in BASIC, which are distributed by the International Ground Water Modeling Center under the name BEAVERSOFT. This is a package of analytical and numerical solutions for ground-water flow and solute transport problems, including steady and unsteady two-dimensional flow in nonhomogeneous aquifers; flow through dams; transport of pollution by advection and dispersion; and saltwater intrusion. The computer programs serve educational purposes in illustrating the material presented in the book but are sufficiently advanced and generalized to provide tools for some professional applications as well.

Anderson and Woessner (1992) discuss selected widely used models in their book as well as principles and practices of ground-water modeling in general. The selected models are MODFLOW, PLASM, and AQUIFEM-1. MODFLOW and PLASM are addressed later in this chapter.

United Nations Modeling Package

A series of ground-water programs is available from the United Nations Department of Technical Co-operation for Development, Natural Resources and Energy Division, Water Resources Branch (1 U.N. Plaza, New York, NY 10017, 212/963-8588).

The MS-DOS-based microcomputer software package (United Nations 1989) includes the following programs:

GW1   Hydraulic Conductivity
GW2   Ground Water Chemistry
GW3   Pumping Tests
GW4   Well Hydraulics and Well Construction
GW5   Water Level Data Base and Hydrographs
GW6   Well Logs and Lithological Cross-Sections
GW7   Confined Aquifer Mathematical Model
GW8   Unconfined Aquifer Mathematical Model
GW9   Small Island Mathematical Model
GW10 Two-Layered Aquifer (Data Analysis and Presentation)
GW11 Graphics

U.S. Geological Survey Ground-Water Models

The U.S. Geological Survey has played a particularly notable role in developing ground-water models. USGS models are widely applied both within and outside the agency. Appel and Reilly (1988) cite reports and present summaries for each of 33 USGS ground-water models categorized as follows:

- Ten saturated flow models
- One variably saturated flow model
- Four saturated, solute transport models
- Two saturated, solute, and heat transport models
- Six saturated, freshwater-saltwater models
- Three heat transport models
- One saturated and unsaturated solute or heat transport model
- One aquifer management model
- One chemical equilibrium model
- One chemical mass transfer model
- One characterization of natural waters model

For most of these 33 USGS ground-water models, Appel and Reilly (1988) cite several versions of the models which have been developed and published. Several USGS models are discussed later in this chapter.

EPA Center for Subsurface Modeling Support

As discussed in Chapter 2, the Robert S. Kerr Environmental Research Laboratory serves as the Environmental Protection Agency's Center for Ground-Water Research. The Center for Subsurface Modeling Support (CSMoS) is the component of the Kerr Laboratory which distributes and supports models developed by the Laboratory. Models available from the CSMoS are listed in Table 2.4.

International Ground Water Modeling Center

The clearinghouse for ground-water modeling software operated by the International Ground Water Modeling Center (IGWMC) is also discussed in Chapter 2. The IGWMC distributes and supports models developed by the USGS and EPA as well as by various other agencies, universities, and private firms. Ground-water modeling represents a somewhat unique category of water management models compared to the categories covered in the other chapters, in that a single entity maintains a comprehensive model inventory.

Van der Heijde et al. (1985, 1988) report on the status of ground-water modeling and summarize the models included in the IGWMC database as well models available from a variety of other sources. Van der Heijde and Elnawawy (1993) provide a comprehensive compilation of ground-water models available from various sources, including the IGWMC.

A software catalog is available upon request from the IGWMC which provides a summary of capabilities and other pertinent information for each of the models distributed by the IGWMC. All of the models included in the IGWMC information database cited earlier are not necessarily distributed by the IGWMC. The catalog provides instructions for requesting software, publications, short-course schedules, and the IGWMC newsletter. The ground-water models included in the June 1993 IGWMC software catalog are tabulated in Table 6.1. With the exception of the special order software, a nominal handling fee varying from $50 to $200 is charged for each model. The fees for the few special order models are significantly higher.

Discussion of Several Representative State-of-the-Art Models

The MODFLOW, PLASM, RANDOM WALK, MOC, WHPA, SUTRA, and SWIFT models are included in the Model Inventory Appendix. These models are representative of the state of the art, include some of the most widely used of the available models, and cover a broad range of modeling capabilities. These and several related models are discussed in the following paragraphs. As is true for the model categories covered in all of the other chapters, citing of specific software does not imply

TABLE 6.1   Models Available from the
International Ground Water Modeling Center

Saturated flow models	Saturated transport models	Parameter estimation models
BEAVERSOFT	AGU-10	COVAR
CAPZONE	ASM	OPTP/PTEST
GWFLOW	AT123D	PUMPTEST
JBD2D/3D	BEAVERSOFT	TETRA
MICROFEM	BIOPLUME	TGUESS
MODFLOW	CANVAS	THCVFIT
PAT	EPA-VHS	THEISFIT
PREMOC	HST3D	TIMELAG
PLASM	MAP	TSSLEAK
RADFLOW	MICROFEM	VARQ
SWAMFLOW	MOC	ONESTEP
THWELLS	MOCDENSE	SOIL
WHPA	PLUME2D	CATTI
ZONEBUDGET	RANDOM WALK	CFITIM
	RWH	CXTFIT
Unsaturated flow models	SOLUTE	WELL
BEAVERSOFT	SUMMERS	
HYDRUS	SUTRA	*Statistical analysis models*
INFIL	SWICHA	GEO-EAS
RETC	VIRALT	GEOPACK
SWACROP		MAP
SWMS 2D	*Unsaturated transport models*	
		Historic models
Hydrogeochemical models	BEAVERSOFT	
	CANVAS	BALANCE
MINTEQA2	CHEMFLO	FP
NETPATH	HYDRUS	PLUME
PHREEQE	ONED	SOHYP
PHRQPITZ	PESTAN	ST2D
WATEQ4F	PESTRUN	SUMATRA-1
	RITZ	TRAFRAP
Special order software	SWMS 2D	UNSAT1
ARMOS	VLEACH	USGS-2D-FLOW
MOTRANS		USGS-3D-FLOW
SOILPROP		
VENTING		

SOURCE: June 1993 IGWMC software catalog.

endorsement over other models. Other excellent ground-water computer programs are available. Practitioners should consider the entire spectrum of available models for their particular applications. However, these highlighted programs provide an excellent starting point for investigating available capabilities.

MODFLOW. The U.S. Geological Survey MODFLOW is a modular, three-dimensional, finite difference ground-water flow model developed by McDonald and Harbaugh (1988). MODFLOW is probably the most widely used of all the available ground-water flow models. Recently updated versions of MODFLOW and associated preprocessor and postprocessor software are available from the USGS, IGWMC (International Ground Water Modeling Center 1993), Scientific Software Group (Scientific Software Group 1993), and others. MODFLOW is documented by McDonald and Harbaugh (1988). A MODFLOW instructional manual and example data sets are available from the IGWMC. Anderson and Woessner (1992) include discussions of the model. Walton (1992) discusses the use of various utility programs along with MODFLOW. The USGS has an inverse model based on MODFLOW, called MODFLOWP, for parameter estimation (Hill 1990). A number of other models related to MODFLOW have been developed.

The MODFLOW modular structure consists of a main program and series of highly independent subroutines called modules. The modules are grouped into packages. Each package deals with a specific feature of the hydrologic system or with a specific method for solving the governing equations. The modular structure facilitates development of additional capabilities because new modules or packages can be added to the program without modifying the existing modules or packages. The input and output systems of the computer program are also designed to permit flexibility.

MODFLOW simulates two-dimensional areal or cross-sectional and quasi- or fully three-dimensional steady or transient saturated flow in anisotropic, heterogeneous, layered aquifer systems. Layers may be simulated as confined, unconfined, or convertible between the two conditions. The model allows for analysis of external influences such as wells, areal recharge, drains, evapotranspiration, and streams. The model incorporates a block-centered, finite difference approach. The finite difference equations are solved by either the strongly implicit procedure or the slice-successive overrelaxation procedure.

PLASM. The Prickett-Lonnquist Aquifer Simulation Model (PLASM) was one of the first readily available, well-documented ground-water flow models. The original model developed by the Illinois State Water Survey (Prickett and Lonnquist 1971) has been updated to a user-friendly format for execution on IBM-compatible microcomputers. Several versions and extensions of the original PLASM are available, including the RANDOM WALK and GWFL3D models (Walton 1989b) discussed later. A discussion of PLASM is included in Anderson and Woessner (1992).

PLASM provides capabilities for simulating two-dimensional unsteady flow in heterogeneous anisotropic aquifers under water table, nonleaky, and leaky artesian conditions. The model allows representation of time varying pumpage from wells, natural or artificial recharge rates, the relationships of water exchange between surface waters and the ground-water reservoir, the process of ground-water evapotranspiration, and the mechanism of converting from artesian to water table conditions. PLASM incorporates an iterative, alternating direction, implicit finite difference solution of the equations of ground-water flow.

RANDOM WALK.   Prickett et al. (1981) incorporated PLASM as a component of a "random walk" solute transport model published by the Illinois State Water Survey. Coupled, the PLASM flow and RANDOM WALK solute transport models provide the aforementioned PLASM flow modeling capabilities. The RANDOM WALK model simulates one- or two-dimensional contaminant transport employing discrete parcel random walk techniques. Contaminant transport is based on a particle in a cell technique for advective mechanisms and a random walk technique for dispersion mechanisms. The effects of convection, dispersion, and chemical reactions are included. The solute transport model simulates continuous and slug contaminant source areas of various shapes, contaminant sinks such as wells and streams, vertically averaged saltwater fronts, and contaminant leakage from overlying source beds.

GWFL3D and GWTR3D.   Walton (1989b) introduced two models, called GWFL3D and GWTR3D, which are modifications and extensions to the PLASM and RANDOM WALK models. The GWFL3D model reflects modifications to PLASM to accommodate quasi-three-dimensional simulations involving multiple stacks of aquifers and confining beds. Unlike RANDOM WALK, GWTR3D decouples the flow model from the solute transport model, calls for a head database, and simulates one time increment per program run.

MOC.   The U.S. Geological Survey Method of Characteristics (MOC) model of two-dimensional solute transport and dispersion was originally developed by Konikow and Bredehoeft (1978). Updated versions of the model and associated preprocessor and postprocessor software are available from the USGS, International Ground Water Modeling Center, Scientific Software Group, and others.

MOC is a two-dimensional model for the simulation of nonconservative solute transport in saturated ground-water systems. MOC is generalized and flexible for application to a wide range of problems. It computes changes in the spatial concentration distribution over time caused by convective transport, hydrodynamic dispersion, mixing or dilution from recharge, and chemical reactions. The chemical reactions include first-order irreversible rate reactions (such as radioactive decay);

reversible equilibrium-controlled sorption with linear, Freundlich, or Langmuir iso-therms; and reversible equilibrium-controlled ion exchange for monovalent or di-valent ions. The model assumes that fluid density variations, viscosity changes, and temperature gradients do not affect the velocity changes, and temperature gradients do not affect the velocity distribution. MOC allows modeling heterogeneous and anisotropic (confined) aquifers.

MOC solves the ground-water flow equation and the nonconservative solute transport equation in a stepwise (uncoupled) fashion. The computer program uses the alternating direction implicit method or the strongly implicit procedure to solve the finite difference approximation of the ground-water flow equation. The strongly implicit procedure for solving the ground-water flow equation is most useful when areal discontinuities in transmissivity exist or when the alternating direction implicit solution does not converge. The MOC model uses the method of characteristics to solve the solute transport equation. It uses a particle tracking procedure to represent convective transport and a two-step explicit procedure to solve the finite difference equation that describes the effects of hydrodynamic dispersion, fluid sources and sinks, and divergence of velocity. The explicit procedure is subject to stability criteria, but the program automatically determines and implements the time step limitations necessary to satisfy the stability criteria. MOC uses a rectangular, block-centered, finite difference grid for flux and transport calculations. The program allows spatially varying diffuse recharge or discharge, saturated thickness, transmissivity, boundary conditions, initial heads and initial concentrations, and an unlimited number of injection or withdrawal wells.

WHPA. The Wellhead Protection Area (WHPA) model was developed by the Environmental Protection Agency originally as a tool for federal, state, and local entities in the delineation of Wellhead Protection Areas as defined by the 1986 Amendments to the Safe Drinking Water Act, but the model can be applied to many different types of problems associated with wells. The model delineates capture zones and contaminant fronts assuming steady-state horizontal flow in the aquifer. The microcomputer modeling system includes a menu-driven user interface.

WHPA consists of four different particle tracking modules: RESSQC, MWCAP, GPTRAC, and MONTEC. The RESSQC module delineates time-related capture zones around multiple pumping wells or contaminant fronts around multiple injection wells in homogeneous aquifers of infinite areal extent with steady and uniform ambient ground-water flow. RESSQC accounts for multiple-well interfer-ence effects. MWCAP is similar to RESSQC but can incorporate stream or barrier boundary conditions for semi-infinite aquifers. It can be used to delineate steady, time-related, or hybrid capture zones. The GPTRAC module contains two options: (1) semianalytical and (2) numerical. The semianalytical option is similar to RESSQC and MWCAP but can accommodate a wider range of aquifer and boundary conditions. This option can simulate delineation in homogenous confined, leaky-confined, or unconfined aquifers with areal extent. The aquifer may be of infinite areal extent or may be bounded by one or two parallel stream and/or barrier (semipermeable with ambient flow) boundaries. The numerical option performs particle tracking using a head field obtained from a numerical ground-water flow model, such as MODFLOW, and accounts for many types of boundary conditions as well as aquifer heterogeneities and anisotropies. MONTEC, the fourth module, performs uncertainty analyses, based on Monte Carlo techniques, for time-related capture zones for a single pumping well in a confined and leaky confined homogeneous aquifer of infinite areal extent.

WHPA is useful as a screening tool in delineating capture zones or contaminant fronts because it requires relatively few parameters. The general parameters required by all the WHPA modules are transmissivity, porosity, saturated thickness, and the rates of recharge or discharge from the simulated injection or pumping wells. Other parameters required for specific modules include location and type of boundaries, the areal recharge rate, confining bed hydraulic conductivity, and thickness of the confining bed.

WHPA presents the modeling results as a plot of the capture zone and particle paths. When simulating multiple wells, the respective capture zones and particle paths are shown in different colors. Plots of up to 15 simulations can be overlayed, one on top of the other, for comparison analysis. A hard copy of the plot as well as a tabulation of the data can be printed using most standard printers or plotters. Plot files can be transported in ASCII format as input to ARC/INFO or other GIS software.

SUTRA.   The Saturated-Unsaturated Transport (SUTRA) model was developed by the U.S. Geological Survey (Voss 1984). SUTRA simulates fluid movement and solute and energy transport in a subsurface environment. The model computes and outputs fluid pressures, solute concentrations, and temperatures as a function of time and location. Options are also available for printing fluid velocities within the system and fluid mass and solute mass or energy budgets for the system. Capabilities are provided for areal and cross-sectional modeling of saturated ground-water flow systems and for cross-sectional modeling of unsaturated zone flow. Boundary conditions, sources, and sinks may be time dependent. The model employs a two-dimensional, hybrid finite element and integrated finite difference solution of the governing equations.

SUTRA may be used to analyze a variety of ground-water contaminant transport and aquifer restoration problems. The model simulates natural or human-induced chemical species transport, including processes of solute sorption, production, and decay. Solute transport modeling may include variable-density leachate movement. Saltwater intrusion in aquifers at near-well or regional scales may be modeled with either dispersed or relatively sharp transition zones between fresh-water and saltwater. SUTRA energy transport simulation capabilities may be employed to model thermal regimes in aquifers, subsurface heat conduction, aquifer thermal energy storage systems, geothermal reservoirs, thermal pollution of aquifers, and natural hydrogeologic convection systems. Past applications of SUTRA include those reported by Voss and Souza (1987) and Bush (1988).

SWIFT-II. The Sandia Waste Isolation, Flow, and Transport (SWIFT-II) model simulates flow and transport processes in both fractured and porous media. SWIFT-II was developed for use in the analysis of deep geologic nuclear waste disposal facilities. However, the generalized model is equally applicable to other problem areas, such as injection of industrial wastes into saline aquifers; heat storage in aquifers; *in situ* solution mining; migration of contaminants from landfills; disposal of municipal wastes; saltwater intrusion in coastal regions; and brine disposal from petroleum storage facilities.

SWIFT-II was developed by the Sandia National Laboratories, U.S. Department of Energy, for the U.S. Nuclear Regulatory Commission. The model is documented by a set of three reports (Reeves et al. 1986a,b,c). SWIFT evolved from the U.S. Geological Survey Waste Injection Program (SWIP) model (INTERCOMP 1976).

SWIFT-II is a transient, three-dimensional model applicable to geologic media which may be fractured, and it solves coupled equations for flow and transport. The processes considered are (1) fluid flow, (2) heat transport, (3) dominant-species (brine) miscible displacement, and (4) trace-species (radionuclides) miscible displacement. The first three processes are coupled via fluid density and viscosity. Together they provide the velocity field required in the third and fourth processes. Both dual-porosity and discrete-fracture conceptualizations may be considered for the fractured zones. A variety of options are provided to facilitate various uses of the model.

Other models. The U.S. Geological Survey Heat and Solute Transport in Three-Dimensional Ground-Water Flow Systems (HST3D) model (Kipp 1987) provides comparable capabilities as SWIFT-II for porous media. The FTWORK model (Faust et al. 1993), developed by GeoTrans, Inc. (Sterling, Virginia, 703/444-7000), is another three-dimensional flow and solute transport model which is less comprehensive but simpler than SWIFT-II. FTWORK contains an inverse routine for calibration of steady-state flow problems.

Multiphase flow may involve either miscible fluids, which mix and combine readily, or immiscible fluids, which do not mix with water. Miscible fluids may be simulated with the previously noted solute transport models. Immiscible fluids, such as nonaqueous-phase liquids (NAPLs), are much more complex to model. SWAN-

FLOW (Faust et al. 1989) is a three-dimensional, finite difference code for simulating the flow of water and an immiscible nonaqueous phase under saturated and unsaturated near-surface conditions. SWANFLOW is available from GeoTrans and the IGWMC. TOUGH (Pruess 1987) is a multiphase fracture-flow code that simulates the coupled three-dimensional transport of liquid water, water vapor, air, and heat in variably saturated fractured media or continuous porous media. TOUGH is available from the Energy Science and Technology Software Center (Oak Ridge, Tennessee, 615/576-2606).

Geochemical codes compute the concentrations of ions in solution in water at chemical equilibrium based on a mass balance. Models such as PHREEQE (Parkhurst et al. 1980) can also simulate the change in speciation that will occur with the addition or removal of a chemical. This procedure is called reaction path modeling. PHREEQE is available from the USGS and IGWMC. Practical application of geochemical codes in contaminant transport modeling has been limited to date.

COMPARISON OF MODELS INCLUDED IN THE MODEL INVENTORY APPENDIX

The following ground-water models are included in the Model Inventory Appendix:

> MODFLOW: Modular Three-Dimensional Finite-Difference Ground Water Flow Model—U.S. Geological Survey
>
> PLASM: Prickett-Lonnquist Aquifer Simulation Model—Illinois State Water Survey
>
> RANDOM WALK: Solute Transport Model—Illinois State Water Survey
>
> MOC: Method of Characteristics Two-Dimensional Solute Transport Model—U.S. Geological Survey
>
> WHPA: Wellhead Protection Area Model—U.S. Environmental Protection Agency
>
> SUTRA: Saturated-Unsaturated Transport Model—U.S. Geological Survey
>
> SWIFT: Sandia Waste Isolation, Flow, and Transport Model—Sandia National Laboratories.

These are all reasonably well-documented, generalized software packages with executable versions available for MS-DOS-based microcomputers. The FORTRAN programs have been compiled and run on other computer systems as well. Various versions and variations of the models are available along with associated preprocessor and postprocessor programs. The MODFLOW and PLASM flow models have been the most extensively applied of all the models. RANDOM WALK and MOC have probably been the most widely used of the five solute transport models.

The seven software packages provide capabilities for simulating a broad range of ground-water flow and water quality processes. The models can be compared based on the processes which can be simulated and associated modeling assumptions. Basic considerations in formulating a modeling approach for a particular application include the following questions:

- Does the application involve only water quantity (flow) considerations or is water quality also of concern?
- If water quality is of concern, what physical, chemical, and/or biological transformation and transport processes are important for the particular application?
- Is only saturated porous media being modeled or is either unsaturated flow conditions or flow through fractured rock also of concern?
- Recognizing that modeling of the three-dimensional real world is greatly simplified by assuming that variations (fluxes) are limited to less than three dimensions, what spatial dimensions are appropriate to optimally minimize modeling complexity while realistically representing the essential aspects of the problem?
- Can steady-state conditions be assumed, or are transient variations over time important?

PLASM and MODFLOW are two- and three-dimensional, respectively, flow models. The other models include simulation of solute transport as well as flow. RANDOM WALK is a solute transport model which incorporates PLASM flow computations. RANDOM WALK, MOC, and SUTRA provide capabilities for two-dimensional simulations. SWIFT-II is a three-dimensional model. SWIFT-II is designed for simulating flow and transport processes in fractured rock as well as porous media. SUTRA models unsaturated as well as saturated flow. The models simulate transient as well as steady-state conditions. WHPA delineates wellhead capture zones and contaminant fronts.

REFERENCES

ANDERSON, M. P., D. S. WARD, E. G. LAPPALA, and T. A. PRICKETT, "Computer Models for Subsurface Water," *Handbook of Hydrology* (D. R. Maidment, editor), McGraw-Hill, 1993.
ANDERSON, M. P., and W. W. WOESSNER, *Applied Groundwater Modeling, Simulation of Flow and Advective Transport,* Academic Press, 1992.

APPEL, C. A., and T. E. REILLY, "Selected Reports that Include Computer Programs Produced by the U.S. Geological Survey for Simulation of Ground-Water Flow and Quality," Water-Resources Investigations Report 87-4271, U.S. Geological Survey, Reston, Virginia, 1988.

ARAL, M. M., *Ground Water Modeling in Multilayer Aquifers,* Lewis Publishers, 1990.

BEAR, J., and A. VERRUIJT, *Modeling Groundwater Flow and Pollution,* D. Reidel Publishing Company, 1987.

BEAR, J., and Y. BACHMAT, *Introduction to Modeling of Transport Phenomena in Porous Media,* Kluwer Academic Publishers, 1990.

BONN, B., and S. ROUNDS, *DREAM: Analytical Groundwater Flow Programs,* Lewis Publishers, 1990.

BUSH, P. W., "Simulation of Saltwater Movement in the Floridan Aquifer System, Hilton Head Island, South Carolina," Water Supply Paper 2331, U.S. Geological Survey, Reston, Virginia, 1988.

CLARKE, D., *Microcomputer Programs for Groundwater Studies,* Developments in Water Science Series, No. 30, Elsevier Science Publishers, 1987.

DOMENICO, P. A., and F. W. SCHWARTZ, *Physical and Chemical Hydrogeology,* John Wiley & Sons, 1990.

FAUST, C. R., J. H. GUSWA, and J. W. MERCER, "Simulation of Three-Dimensional Flow of Immiscible Fluids within and below the Unsaturated Zone," *Water Resources Research,* Vol. 25, No. 12, 1989.

FAUST, C. R., P. N. SIMS, C. P. SPALDING, P. F. ANDERSEN, B. H. LESTER, M. G. SHUPE, and A. HARROVER, "FTWORK: Groundwater Flow and Solute Transport in Three Dimensions, Documentation Version 2.8," GeoTrans, Inc., Sterling, Virginia, March 1993.

FETTER, C. W., *Contaminant Hydrogeology,* Macmillan, 1993.

FREEZE, R. A., and J. A. CHERRY, *Groundwater,* Prentice Hall, 1979.

Geraghty and Miller, Inc., "Geraghty & Miller's Groundwater Bibliography," 5th edition, Water Information Center, Inc., Syosset, New York, 1991.

HELWEG, O. J., *Microcomputer Applications in Water Resources,* Prentice Hall, 1991.

HILL, M. C., "MODFLOWP: A Computer Program for Estimating Parameters of a Transient, Three-Dimensional Groundwater Flow Model Using Nonlinear Regression," Open-File Report 91-484, U.S. Geological Survey, 1990.

INTERCOMP RESOURCE DEVELOPMENT AND ENGINEERING, INC., "Development of Model for Calculating Disposal in Deep Saline Aquifers," USGS/WRI-76-61, PB-256903, National Technical Information Service, 1976.

INTERNATIONAL GROUND WATER MODELING CENTER, "IGWMC Software Catalog," Colorado School of Mines, Golden, Colorado, June 1993.

ISTOK, J., *Groundwater Modeling by the Finite Element Method,* Water Resources Monograph 13, American Geophysical Union, Washington, D.C., 1989.

JAVANDEL, I., C. DOUGHTY, and C. F. TSANG, *Groundwater Transport: Handbook of Mathematical Models,* Water Resources Monograph 10, American Geophysical Union, Washington, D.C., 1984.

KIPP, K. L., "HST3D: A Computer Code for Simulation of Heat and Solute Transport in Three-Dimensional Ground-Water Flow Systems," Water Resources Investigations Report 86-4095, U.S. Geological Survey, Reston, Virginia, 1987.

KONIKOW, L. F., and J. D. BREDEHOEFT, "Computer Model of Two-Dimensional Solute Transport and Dispersion in Ground Water," Techniques of Water-Resource Investigations, Chapter C2, Book 7, U.S. Geological Survey, Reston, Virginia, 1978.

MCDONALD, M. G., and A. W. HARBAUGH, "A Modular Three-Dimensional Finite-Difference Ground-Water Flow Model," Techniques of Water-Resources Investigations 06-A1, U.S. Geological Survey, Reston, Virginia, 1988.

NATIONAL RESEARCH COUNCIL COMMITTEE ON GROUND WATER MODELING ASSESSMENT, Ground Water Models: Scientific and Regulatory Applications, National Academy Press, Washington, D.C., 1990.

PARKHURST, D. L., D. C. THORSTENSON, and L. N. PLUMMER, "PHREEQE—A Computer Program for Geochemical Calculations," Water-Resources Investigations Report 80-96, U.S. Geological Survey, 1980.

PRICKETT, T. A., and C. G. LONNQUIST, "Selected Digital Computer Techniques for Ground-water Resource Evaluation," Bulletin 55, Illinois State Water Survey, Urbana, Illinois, 1971.

PRICKETT, T. A., T. G. NAYMIK, and C. G. LONNQUIST, "A Random-Walk Solute Transport Model for Selected Groundwater Quality Evaluations," Bulletin 65, Illinois State Water Survey, Urbana, Illinois, 1981.

PRUESS, K., "TOUGH User's Guide," NUREG/CR-4645, SAND86-7104, LBL-20700, U.S. Nuclear Regulatory Commission, 1987.

REEVES, M., D. S. WARD, N. D. JOHNS, and R. M. CRANWELL, "Data Input Guide for SWIFT II, the Sandia Waste-Isolation Flow and Transport Model for Fractured Media," NUREG/CR-3162, SAND83-0242, Sandia National Laboratories, Albuquerque, New Mexico, 1986a.

REEVES, M., D. S. WARD, N. D. JOHNS, and R. M. CRANWELL, "Theory and Implementation for SWIFT II, the Sandia Waste-Isolation Flow and Transport Model for Fractured Media," NUREG/CR-3328, SAND83-1159, Sandia National Laboratories, Albuquerque, New Mexico, 1986b.

REEVES, M., D. S. WARD, P. A. DAVIS, and E. J. BONANO, "SWIFT II Self-Teaching Curriculum: Illustrative Problems for the Sandia Waste-Isolation Flow and Transport Model for Fractured Media," NUREG/CR-3925, SAND84-1586, SANDIA National Laboratories, Albuquerque, New Mexico, 1986c.

SCIENTIFIC SOFTWARE GROUP, "Environmental, Engineering, Water Resources Software and Publications, 1993–1994," Washington, D.C., 1993.

United Nations, "Ground Water Software, User's Manual," UN Department of Technical Cooperation for Development, Division of Natural Resources and Energy, Water Resources Branch, New York, December 1989.

VAN DER HEIJDE, P., Y. BACHMAT, J. BREDHOEFT, B. ANDREWS, D. HOLTZ, and S. SEBASTIAN, Groundwater Management: the Use of Numerical Models, 2nd edition, Water Resources Monograph 5, American Geophysical Union, Washington, D.C., 1985.

VAN DER HEIJDE, P. K. M., A. I. EL-KADI, S. A. WILLIAMS, "Groundwater Modeling: An Overview and Status Report," EPA/600/2-89/028, U.S. Environmental Protection Agency, December 1988.

VAN DER HEIJDE, P. K. M., and O. A. ELNAWAWY, "Compilation of Groundwater Models," EPA/600/R-93/118, U.S. Environmental Protection Agency, May 1993.

Voss, C. I., "A Finite-Element Simulation Model for Saturated-Unsaturated, Fluid-Density-Dependent Groundwater Flow with Energy Transport or Chemically-Reactive Single-Species Solute Transport," Water-Resources Investigations Report 84-4369, U.S. Geological Survey, Reston, Virginia, 1984.

Voss, C. I., and W. R. SOUZA, "Variable Density Flow and Solute Transport Simulation of Regional Aquifers Containing a Narrow Freshwater-Saltwater Transition Zone," *Water Resources Research,* American Geophysical Union, Vol. 23, No. 10, 1987.

WALTON, W. C., *Analytical Groundwater Modeling, Flow and Contaminant Migration,* Lewis Publishers, 1989a.

WALTON, W. C., *Numerical Groundwater Modeling, Flow and Contaminant Migration,* Lewis Publishers, 1989b.

WALTON, W. C., *Groundwater Modeling Utilities,* Lewis Publishers, 1992.

YEH, W. W-G., "Systems Analysis in Ground-Water Planning and Management," *Journal of Water Resources Planning and Management,* American Society of Civil Engineers, Vol. 118, No. 3, May/June 1992.

7

WATERSHED RUNOFF MODELS

INTRODUCTION

Watershed models simulate the hydrologic processes by which precipitation is converted to streamflow. The watershed is the system being modeled; precipitation is provided as input; and the runoff characteristics are computed. Water quality is changed during these hydrologic processes. Some models consider only water quantities, whereas others simulate both water quality and quantity. Simulation results essentially always include streamflow hydrographs and sometimes include the associated pollutographs as well. The watersheds being modeled include streams, reservoirs, drainage improvements, and stormwater management facilities as well as the land and land cover on which the precipitation falls.

Model Applications

Watershed models are used to develop streamflow hydrographs required as input for the stream hydraulics models cited in Chapter 8 and the reservoir/river system operation models of Chapter 10. Watershed models may provide both volumetric inflows and pollutant loadings required as input for the river and reservoir water quality models of Chapter 9.

Streamflow hydrographs and/or associated pollutant concentrations are basic data required in many different types of water management modeling applications.

Design hydrographs provide a basis for sizing hydraulic structures such as dams, spillways, flood control improvements, storm sewers, detention basins, culverts, and bridges. Hydrographs are required to delineate flood plains in support of flood-plain management programs. Runoff hydrographs are input to models used to support real-time reservoir system operating decisions. Watershed models are used to quantify the impacts of land use changes and management plans on runoff quantity and quality. Pollutant loading estimates are needed for various water quality management activities. Both urban stormwater management and control of pollution from agricultural activities involve application of watershed models with quality analysis capabilities.

Measured streamflow and contaminant concentration data are often used for the aforementioned types of applications, without the need for watershed models. Historical gauged streamflow and water quality data are typically used in modeling studies to the extent possible. However, watershed models, in combination with precipitation data, are often used to synthesize flows and pollutant loads for the following reasons:

- The availability of streamflow and water quality data is contingent on gauging and sampling stations having been maintained at pertinent locations over a significantly long period of time to develop an adequate database.
- Historical streamflow and water quality data may be nonhomogeneous and not representative of present and future watershed conditions if significant changes in runoff characteristics have occurred because of urbanization or other land use modifications, construction of reservoirs, variations in water supply withdrawals or treatment plant discharges, or other changed conditions during the period of record.
- Modeling studies often involve evaluation of proposed management plans which will affect the runoff quantity and quality characteristics of a watershed.

Watershed Precipitation-Runoff Processes

Some precipitation is loss through the natural hydrologic processes of interception, depression storage, infiltration, evaporation, and transpiration. The remaining precipitation flows overland and through the soil, collects as flow in swales and small channels, and eventually becomes runoff to streams. Ground water also contributes to streamflow, largely independently of the particular precipitation-runoff event. Contaminants enter the water during the runoff processes. Various pollutant transport and transformation processes occur within the hydrologic processes. Land use, drainage improvements, storage facilities, and other development activities significantly affect the processes by which precipitation is converted to streamflow. Snowfall and snowmelt as well as rainfall are important in many areas.

Watershed modeling involves computing flow rates and sometimes contaminant concentrations or loads, as a function of time, at the watershed outlet (or multiple subwatershed outlets) for specified precipitation input. Larger watersheds are typically divided into a number of smaller more hydrologically homogeneous subwatersheds for modeling purposes. The runoff hydrographs from the individual subwatersheds are routed through stream reaches and combined at appropriate locations. Runoff from subwatersheds may also be routed through storm sewer systems and temporarily stored in reservoirs and detention facilities.

Watershed Modeling Literature

A number of hydrology textbooks cover the fundamentals of watershed (precipitation-runoff) modeling, including Linsley et al. (1982), Chow et al. (1988), Viessman et al. (1989), McCuen (1989), Ponce (1989), Singh (1992), and Bedient and Huber (1992). An American Society of Agricultural Engineers monograph edited by Hann, et al. (1982) also provides a thorough coverage of watershed modeling. Several of these references inventory available generalized operational watershed models. Although water quality is addressed by several of the aforementioned references, the primary focus is on water quantity. Wanielista and Yousef (1993) and James (1993) emphasize water quality considerations. DeVries and Hromadka (1993) review a number of surface water quantity and/or quality models. The American Society of Civil Engineers Task Committee on Definition of Criteria for Evaluation of Watershed Models (1993) addresses the problems practicing engineers face in attempting to evaluate the usefulness of watershed models.

Model Categorization

Watershed models can be categorized as single event or continuous. Single-event models are designed to simulate individual storm events and have no capabilities for the soil infiltration capacity and other watershed abstraction capacities to be replenished during extended dry periods. Continuous models simulate long periods of time which include multiple precipitation events separated by significant dry periods with no precipitation. Some models can be used optionally in either single-event or continuous modes.

Many watershed models are designed for quantity-only applications and contain no features for modeling water quality. Other models provide capabilities for analyzing the quality as well as quantity of runoff. Although individual precipitation events can be simulated, most water quality models provide capabilities for continuous modeling.

The following review of available watershed models is organized based on dividing models into three groups:

- Event models without water quality simulation capabilities
- Continuous models without water quality simulation capabilities
- Watershed models with water quality simulation capabilities

REVIEW OF AVAILABLE MODELS

Event Models without Water Quality Modeling Capabilities

Many of the major computer models for simulating the runoff response of single rainfall events, without considering water quality, are based on the unit hydrograph modeling approach. An alternative common modeling approach, often associated with urban stormwater runoff models, utilizes kinematic wave watershed routing. With either approach, the following tasks are accomplished for each individual subwatershed:

1. A precipitation depth or volume is specified for each computational time interval.
2. The runoff volume, resulting from the precipitation in each time interval, is computed.
3. Either a unit hydrograph or kinematic routing is applied to convert the incremental runoff volumes to a runoff hydrograph at the subwatershed outlet.

A precipitation hyetograph is developed based alternatively on gauged historical or real-time storms, frequency-duration-depth relationships, hypothetical design storms such as the probable maximum storm, and/or snowmelt computations. Various alternative methods are available for computing the runoff volume associated with each incremental precipitation volume. The unit hydrograph, kinematic routing, or other alternative approach is used to convert the volume increments to a direct runoff hydrograph (flow rate versus time) at the watershed outlet. Base flow from ground water, wastewater treatment plant discharges, or other sources, if any, is added to the runoff hydrograph to obtain the total streamflow hydrograph. Streamflow hydrographs at upstream locations are routed through stream reaches and combined with runoff hydrographs from other subwatersheds, as appropriate, to develop streamflow hydrographs at other downstream locations.

HEC-1. The HEC-1 Flood Hydrograph Package is probably the most widely used of the numerous available watershed models. HEC-1 and HEC-2 (covered in Chapter 8) are perhaps the most widely used of all the models cited in this book. HEC-1 and HEC-2 are often used in combination for flood-plain studies. For a particular flood event, peak discharges computed with HEC-1 are inputted to HEC-2, which computes the corresponding water surface profiles. User manuals, training documents, and papers and reports on specific applications are available from the Hydrologic Engineering Center. Several universities, as well as the Hydrologic Engineering Center, conduct annual several-day-long short courses for practicing engineers on application of HEC-1 and HEC-2. HEC-1 and HEC-2 are also used in regular undergraduate and graduate hydrology and hydraulics courses at several universities and are discussed in textbooks such as Bedient and Huber (1992). References on applying HEC-1 and HEC-2, such as the book by Hoggan (1989), are also available. The HEC list of software vendors includes a number of firms and universities that offer various forms of HEC-1 and HEC-2 user assistance.

HEC-1 provides an extensive package of optional computational methods (USACE Hydrologic Engineering Center 1990). Precipitation-runoff modeling represents the central focus of the package, but other related modeling capabilities are provided as well. In addition to the basic watershed modeling capabilities, the HEC-1 package includes several other optional features involving partially automated parameter calibration, multiplan-multiflood analysis, dam safety analysis, economic flood damage analysis, and flood control system optimization.

An HEC-1 precipitation-runoff modeling application typically involves dividing a watershed into a number of subwatersheds. The computational tasks noted earlier are performed for each subwatershed. HEC-1 provides flexible options for developing and/or inputting precipitation data, which may reflect snowfall and snowmelt as well as rainfall. Precipitation volumes are converted to direct runoff volumes using one of the following optional methods: Soil Conservation Service (SCS) curve number method; initial and uniform loss rate; exponential loss rate function; Holtan loss rate function; or Green and Ampt relationship. Runoff hydrographs are computed from the incremental runoff volumes using either the unit hydrograph or kinematic routing options. A unit hydrograph may be input to HEC-1. Alternatively, the model includes options for developing synthetic unit hydrographs using either the Soil Conservation Service, Snyder, or Clark methods. The kinematic wave watershed routing option included in the HEC-1 package is patterned after the Massachusetts Institute of Technology Catchment Model (MITCAT).

Watershed modeling also involves routing hydrographs through stream reaches and reservoirs. HEC-1 uses hydrologic storage routing for reservoirs. The following channel routing options are provided: Muskingum, Muskingum-Cunge, modified Puls, working R and D, average lag, and kinematic wave.

NexGen. The Hydrologic Engineering Center is in the process of developing a next generation (NexGen) of hydrologic engineering software. HEC-1 will be replaced or updated during this process. The NexGen project encompasses several modeling areas, including precipitation-runoff, river hydraulics, reservoir operations, and flood damage analysis. NexGen models will be designed for interactive use in a multitasking environment. The models will include graphical user interface, data management, graphics, and reporting capabilities as well as the computational engines.

TR-20. The single-event watershed model documented by Soil Conservation Service Technical Release (TR) No. 20 (Soil Conservation Service 1982) has been widely used both within and outside the SCS. TR-20 was originally developed by the Hydrology Branch of the SCS in cooperation with the Hydrology Laboratory, Agricultural Research Service. Since its original release in 1965, several modifications and additions have been made by the SCS and others. TR-20 incorporates the procedures described in the SCS *National Engineering Handbook,* Section 4, Hydrology (NEH-4), except for the new attenuation-kinematic (Att-Kin) channel routing procedure, which has replaced the old convex method. TR-20 uses the SCS curve number method and SCS curvilinear dimensionless unit hydrograph to develop the runoff response of watersheds to a rainfall event. Hydrographs are routed through channel reaches using the attenuation-kinematic routing technique. Modified Puls routing is used for reservoirs.

TR-20 is similar to using HEC-1 with certain options selected. TR-20 is much simpler than the HEC-1 package from the perspective of not including as many options. For example, the SCS curve number method is the only method for converting rainfall volumes to runoff volumes in TR-20 but is just one of the five options for converting rainfall volumes to runoff volumes in HEC-1. The SCS dimensionless unit hydrograph in TR-20 is one of three optional synthetic unit hydrographs included in HEC-1.

HYMO. HYMO (Williams and Hann 1973) was developed by the USDA Agricultural Research Service in cooperation with Texas A&M University. HYMO provides a set of commands which the user selects and combines in any sequence to develop a model for a particular watershed. HYMO incorporates the SCS curve number method (like TR-20 and HEC-1), a two-parameter gamma function synthetic unit hydrograph method, variable storage coefficient channel routing (which is somewhat similar to the working R & D option in HEC-1), and hydrologic storage routing for reservoirs (like HEC-1 and TR-20). Unlike HEC-1 and TR-20, HYMO includes an optional routine to compute watershed sediment yields for individual storms using a modified version of the universal soil loss equation.

MILHY. The Military Hydrology (MILHY) model was developed as a part of the Military Hydrology Research Program at the USACE Waterways Experiment Station (Baird and Anderson 1990). MILHY was developed for use by Army terrain teams in forecasting streamflows that would result from a given rainfall event. Rainfall can be input from a synthetic storm or as measured data from precipitation gauges or radar. MILHY incorporates the following methods used in HYMO: SCS curve number method for computing runoff volumes; HYMO unit hydrograph; variable storage coefficient channel routing; and modified Puls hydrologic storage routing for reservoirs. MILHY also includes standard step method water surface profile computations like HEC-2 and the A&M Watershed Model.

A&M Watershed Model. The A&M Watershed Model was originally developed by expanding MILHY, which incorporates basic computational techniques from HYMO. Like MILHY, the A&M Watershed Model can accept radar readings as well as the more conventional gauged rainfall data. Alternatively, synthetic design storms can be developed within the model. A watershed is typically divided into subwatersheds, with the runoff hydrographs computed for each subwatershed being routed and combined as appropriate. The model computes the runoff response to the rainfall event using either the SCS curve number method or Green and Ampt loss rate function options, combined with a two-parameter gamma function unit hydrograph which can be adjusted by urbanization peaking factors. Channel routing options include HYMO variable storage coefficient routing and a recently developed hydraulic routing method. The conventional reservoir storage routing approach is used. The model also includes computation of water surface profiles using the standard step method solution of the energy equation.

The A&M Watershed Model is oriented toward urban drainage and stormwater management. The modeling package includes several optional capabilities for design and analysis of storm sewers, culverts, detention basins, and sedimentation basins. The model also includes an option for performing frequency analyses of inputted annual peak discharges using the log-Pearson type III or Gumbel probability distributions.

The A&M Watershed Model is documented by a user's manual (James et al. 1990). Several journal papers discuss particular features of the model (Garcia and James 1988; James et al. 1987, 1992, 1993; Pridal and James 1989).

Other single-event models. Viessman et al. (1989), Ponce (1989), Singh (1992), and DeVries and Hromadka (1993) cite a number of watershed models including the following single-event models without water quality capabilities: Massachusetts Institute of Technology Catchment Model (MITCAT), Illinois Urban Drainage Area Simulator (ILLUDAS), U.S. Geological Survey Distributed Routing Rainfall-Runoff Model (DR3M), University of Cincinnati Urban Runoff Model, and Penn State Urban Runoff Model. These models are oriented toward small urban watersheds. The MOUSE (Modeling Of Urban SEwers) software package is widely used in Europe, Australia, and New Zealand for analysis of urban drainage problems and includes modules for runoff, pipe flow, and pollution transport (Urbonas and Stahre 1990). The model inventory compiled by the U.S. Bureau of Reclamation (1991) includes the Bureau of Reclamation Flood Hydrograph and Routing (FHAR) model, several models for predicting runoff and soil erosion from small watersheds developed by the Agricultural Research Service, and several other precipitation-runoff models.

Continuous Models without Water Quality Modeling Capabilities

PRMS. The U.S. Geological Survey Precipitation-Runoff Modeling System (PRMS) performs computations on both a daily and smaller time-interval storm scale using variable time steps (DeVries and Hromadka 1993). During a storm event, time intervals as small as a minute may be used. Runoff from a storm event is computed using kinematic flood routing for a watershed represented by interconnected flow planes and channels. The kinematic routing for an individual precipitation event is similar to the USGS Distributed Routing Rainfall-Runoff Model (DR3M) documented by Alley and Smith (1982). A daily interval is used between storm events. Streamflow is computed as a mean daily flow.

The USGS has developed maps of hydrologic response units (HRU) for the United States. Each HRU is assumed to have homogeneous hydrologic characteristics. A watershed is composed of a number of these units. Parameters for a HRU include surface slope, aspect, elevation, soil type, vegetation type, and distribution of precipitation. The PRMS model performs a water balance and energy balance for each HRU. The watershed response is the sum of all pertinent HRU responses.

The PRMS is used in combination with the USGS ANNIE data management program, described in Chapter 3, and a modified version of the National Weather Service (NWS) Extended Streamflow Prediction (ESP) model. PRMS, ANNIE, and ESP used together provide a comprehensive watershed modeling system. Capabilities are provided for compiling and developing the necessary data, simulating and forecasting watershed response, and analyzing model results graphically and statistically.

SHE. The European Hydrologic System or Système Hydrologique Europóen (SHE) was developed jointly by the Danish Hydraulic Institute, United

Kingdom Institute of Hydrology, and SOGREAH in France with financial support from the Commission of European Communities (Abbott et al. 1986; DeVries and Hromadka 1993). SHE is a physically based, distributed-parameter watershed modeling system which incorporates the major hydrologic processes, including precipitation, snowmelt, canopy interception, evapotranspiration, overland flow, saturated and unsaturated subsurface flow, and channel flow. Spatial variability of the hydrologic processes is represented by a rectangular grid in the horizontal plane and vertically by a series of horizontal planes at various depths. SHE may be applied in analyzing irrigation schemes, land use changes, water development projects, groundwater contamination, erosion and sediment transport, and floods.

SSARR. The Streamflow Synthesis and Reservoir Regulation (SSARR) model consists of three basic components: (1) a watershed model for synthesizing runoff from rainfall and snowmelt, (2) a streamflow routing model, and (3) a reservoir regulation model (USACE North Pacific Division 1987). The SSARR reservoir regulation simulation capabilities are noted in Chapter 10. SSARR is a continuous watershed model designed for large river basins. Various versions of SSARR date back to 1956. The USACE North Pacific Division originally developed SSARR as a generalized tool for planning, design, and operation of river basin development projects. The North Pacific Division initially applied the SSARR to operational flow forecasting and river management activities in the Columbia River Basin. The model was later used by the Cooperative Columbia River Forecasting Unit, composed of the USACE, National Weather Service, and Bonneville Power Administration. Numerous other river systems in the United States and other countries also have been modeled by various entities using SSARR.

Streamflow hydrographs are synthesized from rainfall and snowmelt runoff. Rainfall data are provided as input. Snowmelt can be computed based on inputted precipitation depth, elevation, air and dewpoint temperatures, albedo, radiation, and wind speed. Snowmelt options include the temperature index method or the energy budget method. Application of the model begins with a subdivision of the river basin into hydrologically homogeneous subwatersheds. For each subwatershed, the model computes base flow, subsurface or interflow, and surface runoff. Each flow component is delayed according to different processes, and all are then combined to produce the total subwatershed outflow hydrograph. The subwatershed outflow hydrographs are routed through stream reaches and reservoirs and combined with outflow hydrographs from other subwatersheds to obtain streamflow hydrographs at pertinent locations in the river system.

SWM-IV. The Stanford Watershed Model (SWM-IV) and its variations are composed of a set of water budget accounting procedures which incorporate computational routines for the various hydrologic processes, such as interception, infiltration, evapotranspiration, overland flow, channel routing, and so forth. Time series input data include precipitation, potential evapotranspiration, and, if snowmelt is modeled, additional meteorological data. In the model, precipitation is stored in three soil-moisture zones (upper, lower, ground water) and in the snowpack. The upper and lower storage zones account for overland flow, infiltration, interflow, and inflow to ground-water storage. Ground-water storage supplies baseflow to stream channels. Evaporation and transpiration may occur from any of the three storage zones. The runoff from overland flow, interflow, and baseflow enters the channel system and is routed downstream. Model output includes continuous outflow hydrographs.

The Stanford Watershed Model (SWM-IV) was developed at Standard University in the early 1960s (Crawford and Linsley 1966). Several watershed models were developed during the 1970s based on SWM-IV. These other versions of the Stanford Watershed Model include the Kentucky Watershed Model, Texas Watershed Model, Ohio Watershed Model, U.S. Department of Agriculture Hydrograph Laboratory (USDAHL) Model, Sacramento Model, National Weather Service River Forecast System, Hydrocomp Simulation Program, and Hydrological Simulation Program-Fortran (Ponce 1989; Viessman et al. 1989).

Watershed Models with Water Quality Modeling Capabilities

The Hydrocomp Simulation Program (HSP) and Hydrological Simulation Program—Fortran (HSPF) are built on the Stanford Watershed Model (SWM-IV) but include the addition of water quality analysis capabilities. HSP is a commercially available program developed by Hydrocomp, Inc. HSPF is a public domain model originally developed by Hydrocomp for the Environmental Protection Agency, which continues to maintain the model.

HSPF. The HSPF is documented by Johanson et al. (1980, 1984). HSPF provides relatively sophisticated capabilities for continuous simulation of a broad range of hydrologic and water quality processes. The model is oriented more toward agricultural and other nonurban watersheds, but urban watersheds can also be simulated. HSPF consists of a set of modules arranged in a hierarchical framework built around a time series management system. The various simulation and utility modules can be invoked individually or in various combinations. The structured design of the model facilitates users adding their own modules, if they so desire.

HSPF simulates watershed hydrology and water quality for both conventional and toxic organic pollutants. Input data include time histories of rainfall, temperature, and solar radiation and information regarding land surface characteristics (such as land use patterns and soil properties) and land management practices. The result of the simulation of a subwatershed is a hydrograph and pollutographs. The model predicts flow rates, sediment loads, and nutrient and pesticide concentrations. The subwatershed runoff characteristics are then used by the model to simulate instream processes to determine hydrographs and pollutographs at all pertinent locations in the watershed. HSPF allows integrated simulation of land and soil contaminant runoff processes with instream hydraulic and sediment-chemical interactions.

HSPF simulates three sediment types (sand, silt, and clay) in addition to a single organic chemical and transformation products of that chemical. The transfer and reaction processes modeled are hydrolysis, oxidation, photolysis, biodegradation, volatilization, and sorption. Sorption is modeled as a first-order kinetic process in which the user must specify a desorption rate and an equilibrium partition coefficient for each of the three solid types. Benthic exchange is modeled as sorption/desorption and desorption/scour with surficial benthic sediments.

SWMM. The Storm Water Management Model (SWMM) is probably the most widely used of the available urban runoff quantity/quality models. SWMM was originally developed in 1969–1970 by a consortium led by Metcalf and Eddy, Inc., University of Florida, and Water Resources Engineers, Inc., working under the auspices of the Environmental Protection Agency. The model has been continually improved and maintained by the EPA and University of Florida with contributions from various other entities. The EPA coordinates a user group to facilitate interaction and sharing of experiences between the numerous model users. The original SWMM Users Group and the later Storm and Water Quality Model Users Group include participants from throughout the United States and Canada and from other countries as well. The current version of SWMM is documented by Huber and Dickinson (1988). The model is discussed in several hydrology textbooks, including Viessman et al. (1989) and Ponce (1989). James (1993) traces the evolution of SWMM and cites the various conferences and user group meetings which have focused on the model.

The hydrographs and pollutographs developed by SWMM can be used as input to the river and reservoir water quality models described in Chapter 9. In particular, SWMM has linkages to facilitate use with the Environmental Protection Agency's WASP model.

SWMM simulates both water quantity and quality aspects of urban stormwater runoff and combined sewer overflow. SWMM is a large, relatively sophisticated hydrologic, hydraulic, and water quality simulation model. The model provides capabilities for simulating watersheds with complex storm sewer networks, detention storage, and water treatment facilities. The original SWMM was a single-event model, but later versions include capabilities for maintaining water budgets for long-term continuous simulations. Flows are derived from rainfall and/or snow accumulation and melt. Model output includes hydrographs and pollutographs at pertinent locations. The model allows evaluation of storage and treatment facilities by computing and displaying changes in receiving water quality, treatment efficiencies, and associated dollar cost estimates.

Surface runoff is generated based on rainfall hyetographs, antecedent moisture conditions, land use, and topography. Dry-weather sanitary flows are generated based on land use, population density, and other factors. Infiltration into the sewer system is computed based on ground-water levels and sewer condition. Flows and pollutants are routed through the sewer system using a modified kinematic wave approximation, based on the continuity and Manning equations, assuming complete mixing at various inlet points. Routines are provided to modify the hydrographs at selected points in the sewer system to represent storage and/or treatment. Quality constituents handled by SWMM include suspended solids, settleable solids, biochemical oxygen demand (BOD), nitrogen, phosphorus, and grease.

EXTRAN (Roesner et al. 1988) is a stand-alone submodel of SWMM. EXTRAN is a dynamic flow routing model that routes inflow hydrographs through an open channel and/or closed conduit system, computing the time history of flows and heads throughout the system. The routing is based on an explicit numerical solution of the full St. Venant equations.

STORM.   The Storage, Treatment, Overflow, Runoff Model (STORM) is documented by the USACE Hydrologic Engineering Center (1977). STORM simulates the interactions of rainfall/snowmelt, runoff, dry-weather flow, pollutant accumulation and washoff, land surface erosion, treatment, and detention storage. Loads and concentrations are computed for the following basic water quality parameters: suspended and settleable solids, biochemical oxygen demand, total nitrogen, orthophosphate, and total coliform. STORM is a continuous model but can also be used for single events. The model outputs statistical information on quantity and quality of washoff and overflow and pollutographs for selected individual events.

STORM can be used for preliminary sizing of storage and treatment facilities for control of urban stormwater runoff. The impact of land use changes on water quantity and quality can also be evaluated. Pollutographs developed with STORM can be used as input to the receiving water quality models described in Chapter 9.

SWRRB and SWRRB-WQ.   The Simulator for Water Resources in Rural Basins (SWRRB) was developed by the Grassland, Soil, and Water Research Laboratory of the Agricultural Research Service (ARS) of the U.S. Department of Agriculture (Temple, Texas, 817/770-6500). SWRRB is designed to predict the effect of various types of watershed management practices on water and sediment yields in ungaged agricultural watersheds (Arnold et al. 1990; Williams et al. 1985). The major processes reflected in the model include precipitation, surface runoff, percolation, lateral subsurface flow, evapotranspiration, pond and reservoir evaporation, erosion and sedimentation, soil temperature, crop growth, and irrigation. Many years of daily flows may be computed for inputted or computed precipitation data.

Precipitation may be either inputted or developed by the model as a Markov process using inputted probabilities. A watershed may be divided into as many as 10 subwatersheds. The soil profile can be divided into as many as 10 layers. The hydrologic computations are based on the water balance equation. The SCS curve number method is used to compute runoff volumes. Sediment yield is determined using the modified universal soil loss equation and a sediment routing model.

The Simulator for Water Resources in Rural Basins—Water Quality (SWRRB-WQ) was developed by adding water quality modeling capabilities to SWRRB. SWRRB-WQ simulates weather, hydrology, erosion, sediment yield, nitrogen and phosphorus cycling and movement, pesticide fate and movement, crop growth and management, pond and reservoir management, and other processes (Arnold et al. 1992). SWRRB-WQ has been used by the Agricultural Research Service, Soil Conservation Service, Environmental Protection Agency, and other agencies to assess the effects of land management on off-site water quantity and quality, pollution of coastal bays and estuaries, reservoir sedimentation, and registration of pesticides.

Other ARS models. In addition to SWRRB and SWRRB-WQ, the USDA Agricultural Research Service (ARS) has developed several other models for simulating hydrologic and water quality processes in rural watersheds. The recently developed Soil and Water Assessment Tool (SWAT) is designed to extend the capabilities of SWRRB-WQ to large, complex rural river basins. SWAT, like SWRRB, was developed at the ARS Grassland, Soil, and Water Research Laboratory in Temple, Texas. SWAT reflects changes to SWRRB-WQ involving (1) expanding the model to allow simultaneous computations on several hundred subwatersheds and (2) adding components to simulate lateral flow, ground water flow, reach routing transmission losses, and sediment and chemical movement through ponds, reservoirs, streams, and valleys. SWAT is a spatially distributed watershed model that uses a daily time step for simulation periods that may exceed 100 years. Major components of the model include hydrology, weather, sedimentation, soil temperature, crop growth, nutrients, ground water and lateral flow, and agricultural management. SWAT is presently being used in the HUMUS (Hydrologic Unit Model for the United States) project, which entails national and regional water assessments for the major river basins of the United States. SWAT is combined with geographical information systems and relational databases. SWAT is normally run on workstations under the UNIX operating system to utilize the full capabilities of the model. However, an MS-DOS-based microcomputer version is also available.

SWRRB was developed by modifying and expanding the earlier Chemicals, Runoff, and Erosion from Agricultural Management Systems (CREAMS) model, which simulates hydrology, erosion, nutrients, and pesticides from field-size areas (Knisel 1980). SWRRB expands CREAMS for applicability to larger, more complex watersheds. The recently developed Groundwater Loading Effects of Agricultural

Management Systems (GLEAMS) model was designed to replace the earlier CREAMS model (Center for Natural Resource Information Technology 1993). GLEAMS simulates the effects of weather, soils, tillage practices, and pesticide and nutrient management on movement of nutrients, pesticides, and pesticide degradation products to ground and surface waters. GLEAMS is a continuous, field-scale model that permits assessment of the effects of variable topography and slope within the field. The model is used by the USDA, other government agencies, and agricultural chemical companies to assess the environmental effects of alternative management practices and pesticide products.

The Erosion-Productivity Impact Calculator (EPIC) is a comprehensive, field-scale cropping systems model also developed at the ARS Grassland, Soil, and Water Research Laboratory. EPIC uses a daily time step to simulate weather, hydrology, soil erosion, nutrient cycling, crop growth, pesticide fate, and a number of crop management alternatives. EPIC has been used in a number of national and regional analyses, including the 1987 Resources Conservation Appraisal, Farm Bill analyses, prediction of the effects of the 1988 drought, effects of CO_2 and climate change, and the impacts of management practices on water quality in the USDA Water Quality Demonstration Projects.

The Agricultural Non-Point Source (AGNPS) model was developed by the Agricultural Research Service in cooperation with the U.S. Soil Conservation Service and Minnesota Pollution Control Agency (Young et al. 1987, 1989). AGNPS is an event-based rainfall-runoff model for predicting runoff water quality for alternative watershed management practices. Basic model components include hydrology, erosion, sediment transport, and chemical transport. Land surface conditions are represented by a uniform grid of geographic data cells, which may be 1 to 40 acres in size. Runoff and transport processes for sediment, nutrients, and chemical oxygen demand (COD) are simulated for each cell. Flows and pollutants are routed through the channel system to the basin outlet. Point source inputs, such as nutrient COD from animal feedlots, can also be simulated and combined with the nonpoint source pollutants. Runoff volume is computed by the SCS curve number method. An empirical equation is used to determine peak flows. Erosion is estimated using a modified form of the universal soil loss equation. Chemical transport calculations are divided into soluble and sediment-absorbed phases. COD is assumed to be soluble and to accumulate without losses.

COMPARISON OF MODELS INCLUDED IN THE MODEL INVENTORY APPENDIX

The following seven watershed (precipitation-runoff) models are included in the Model Inventory Appendix. Again, in this chapter and all the others, designation of only a limited number of computer programs for inclusion in the Model Inventory

Appendix does not imply that other excellent generalized models are not available. The following models are readily available, widely used generalized programs representative of the state of the art and actual practice.

HEC-1:   Flood Hydrograph Package—Hydrologic Engineering Center

TR-20:   Computer Program for Project Hydrology—Soil Conservation Service

A&M Watershed Model—Texas A&M University

SSARR: Streamflow Synthesis and Reservoir Regulation Model—USACE North Pacific Division

SWMM: Stormwater Management Model—Environmental Protection Agency

HSPF: Hydrologic Simulation Program—Fortran—Environmental Protection Agency

SWRRB-WQ: Simulator for Water Resources in Rural Basins—Water Quality—USDA Agricultural Research Service

HEC-1, TR-20, and the A&M Watershed Model are single-event models without water quality capabilities. TR-20 is the simplest of the models and is limited to basic rainfall-runoff and streamflow routing computations. TR-20 has the advantage of simplicity. Both HEC-1 and the A&M Watershed model include essentially all the capabilities provided by TR-20. Whereas TR-20 provides only one method for each computational task, HEC-1 allows the user to select from several alternative methods. The A&M Watershed also includes alternative computational methods for various tasks, but not to the extent as HEC-1. In addition to the basic rainfall-runoff and streamflow modeling tasks performed by all three models, HEC-1 includes modeling capabilities, not included in the other models, for including snowmelt rather than just rainfall; performing flood control economic analyses; and partially automating parameter calibration. The A&M Watershed model includes capabilities, not provided by HEC-1 and TR-20, for design and analysis of storm sewers, culverts, and detention basins; and for accepting precipitation input data in the form of radar readings. The A&M Watershed model also determines stream water surface profiles using the standard step method solution of the energy equation. The same water surface profile computations can also be performed with HEC-2 (see Chapter 8), which is often used in combination with HEC-1.

SSARR is a continuous model unlike the first three models, which are limited to individual rainfall events. SSARR can be used to synthesize continuous long-term streamflow sequences associated with rainfall and snowmelt. The SSARR water budget accounting computations are somewhat more complex than single-event models and require more input data. SSARR does not include the various optional capabilities included in the HEC-1 and A&M Watershed Model packages. Unlike these other models, SSARR includes capabilities for simulating operation of major reservoir systems.

Unlike the other models, SWMM, HSPF, and SWRRB-WQ contain capabilities for water quality simulation. Although these models can be used for quantity-only applications, they are typically associated with water quality modeling. SWMM is oriented toward urban watersheds. HSPF and SWRRB-WQ are oriented toward rural agricultural watersheds. However, neither of the models is necessarily limited to a particular type of watershed, either urban or rural. These are relatively complex models. A reasonably simple application might involve using only selected options from the extensive package of optional modeling capabilities provided. However, the types of water quality modeling applications typically associated with SWMM, HSPF, and SWRRB-WQ require extensive input data and significant time and effort by experienced, knowledgable modelers.

REFERENCES

ABBOTT, M. B., ET AL., "An Introduction to the European Hydrological System—Système Hydrologique Européen, 'SHE'—1: History and Philosophy of a Physically-Based, Distributed Modelling System, and 2: Structure of a Physically-Based, Distributed Modelling System," *Journal of Hydrology,* Elsevier, Vol. 87, 1986.

ALLEY, W. M., and P. E. SMITH, "Distributed Routing Rainfall-Runoff Model—Version II," Open File Report 82-344, U.S. Geological Survey, Denver, Colorado, 1982.

AMERICAN SOCIETY OF CIVIL ENGINEERS TASK COMMITTEE ON DEFINITION OF CRITERIA FOR EVALUATION OF WATERSHED MODELS, "Criteria for Evaluation of Watershed Models," *Journal of Irrigation and Drainage Engineering,* Vol. 119, No. 3, May/June 1993.

ARNOLD, J. G., J. R. WILLIAMS, A. D. NICKS, and N. B. SAMMONS, "SWRRB, a Basin Scale Simulation Model for Soil and Water Resources Management," Texas A&M University Press, College Station, Texas, 1990.

ARNOLD, J. G., J. R. WILLIAMS, R. H. GRIGGS, and N. B. SAMMONS, "SWRRBWQ, a Basin Scale Simulation Model for Assessing Impacts on Water Quality, Model Documentation and User Manual," Agricultural Research Service, Temple, Texas, 1992.

BAIRD, L., and M. G. ANDERSON, "Flood Inundation Modelling Using MILHY, Final Technical Report, Volume 2, User Manual," U.S. Army Corps of Engineers, European Research Office, London, England, September 1990.

BEDIENT, P. B., and W. C. HUBER, *Hydrology and Floodplain Analysis,* 2nd edition, Addison-Wesley, 1992.

CENTER FOR NATURAL RESOURCE INFORMATION, "Agricultural and Natural Resource Policy Analysis," Texas A&M University System, Blacklands Research Laboratory, Temple, Texas, January 1993.

CHOW, V. T., D. R. MAIDMENT, and L. W. MAYS, *Applied Hydrology,* McGraw-Hill, 1988.

CRAWFORD, N. H., and R. K. LINSLEY, "Digital Simulation in Hydrology: Stanford Watershed Model IV," Technical Report No. 39, Civil Engineering Department, Stanford University, July 1966.

DEVRIES, J. J., and T. V. HROMADKA, "Chapter 21 Computer Models for Surface Water," in *Handbook of Hydrology*, edited by D. R. Maidment, McGraw-Hill, 1993.

GARCIA, A., and W. P. JAMES, "Urban Runoff Simulation Model," *Journal of Water Resources Planning and Management*, American Society of Civil Engineers, Vol. 114, No. 4, July 1988.

HANN, C. T., H. P. JOHNSON, and D. L. BRAKENSIEK (editors), *Hydrologic Modeling of Small Watersheds*, American Society of Agricultural Engineers, 1982.

HOGGAN, D. H., *Computer-Assisted Hydrology & Hydraulics Featuring the U.S. Army Corps of Engineers HEC-1 and HEC-2 Software Systems*, McGraw-Hill, 1989.

HUBER, W. C., and R. E. DICKINSON, "Storm Water Management Model, Version 4: User's Manual," Environmental Research Laboratory, Environmental Protection Agency, Athens, Georgia, August 1988.

JAMES, W., *New Techniques for Modelling the Management of Stormwater Quality Impacts*, Lewis Publishers, 1993.

JAMES, W. P., C. D. ROBINSON, and J. F. BELL, "Radar-Assisted Real-Time Flood Forecasting," *Journal of Water Resources Planning and Management*, American Society of Civil Engineers, Vol. 119, No. 1, January/February 1993.

JAMES, W. P., J. WARINNER, and M. REEDY, "Application of the Green-Ampt Infiltration Equation to Watershed Modeling," *Water Resources Bulletin*, American Water Resources Association, Vol. 28, No. 3, June 1992.

JAMES, W. P., P. H. WINSOR, J. F. BELL, M. G. SPINKS, A. J. GARCIA, D. PRIDAL, J. WARINNER, K. J. KAATZ, V. MORGAN, M. REEDY, and C. ROBINSON, "A&M Watershed Model Users Manual," Texas A&M University, College Station, Texas, January 1990.

JAMES, W. P., P. W. WINSOR, and J. R. WILLIAMS, "Synthetic Unit Hydrograph," *Journal of Water Resources Planning and Management*, American Society of Civil Engineers, Vol. 113, No. 1, January 1987.

JOHANSON, R. C., J. C. IMHOFF, and H. H. DAVIS, "User's Manual for Hydrological Simulation Program—Fortran (HSPF)," EPA-600/9-80-015, Environmental Protection Agency, Environmental Research Laboratory, Athens, Georgia, April 1980.

JOHANSON, R. C., J. C. IMHOFF, J. L. KITTLE, and A. S. DONIGIAN, "Hydrological Simulation Program—Fortran (HSPF): User's Manual for Release 8.0," EPA-600/3-84-066, Environmental Protection Agency, Environmental Research Laboratory, Athens, Georgia, 1984.

KNISEL, W. G. (editor), "CREAMS: A Field Scale Model for Chemicals, Runoff, and Erosion from Agricultural Management Systems," Conservation Research Report No. 26, U.S. Department of Agriculture, 1980.

LINSLEY, R. K., M. A. KOHLER, and J. L. H. Paulhus, *Hydrology for Engineers*, 3rd edition, McGraw-Hill, 1982.

MCCUEN, R. H., *Hydrologic Analysis and Design*, Prentice Hall, 1989.

PONCE, V. M., *Engineering Hydrology*, Prentice Hall, 1989.

PRIDAL, D. B., and W. P. JAMES, "Routing Procedure for Ungaged Channels," *Journal of Water Resources Planning and Management,* American Society of Civil Engineers, Vol. 115, No. 1, January 1989.

ROESNER, L. A., J. A. ALDRICH, and R. E. DICKINSON, "Storm Water Management Model User's Manual Version 4: Extra Addendum," Environmental Research Laboratory, Environmental Protection Agency, Athens, Georgia, August 1988.

SINGH, V. P., *Elementary Hydrology,* Prentice Hall, 1992.

SOIL CONSERVATION SERVICE, "Computer Program for Project Formulation," Technical Release 20, Washington, D.C., 1982.

URBONAS, B., and P. STAHRE, *Stormwater Detention for Drainage, Water Quality, and CSO Management,* Prentice Hall, 1990.

U.S. ARMY CORPS OF ENGINEERS, Hydrologic Engineering Center, "Storage, Treatment, Overflow, Runoff (STORM) Model, User's Manual," August 1977.

U.S. ARMY CORPS OF ENGINEERS, Hydrologic Engineering Center, "HEC-1 Flood Hydrograph Package, User's Manual," CPD-1A, September 1990.

U.S. ARMY CORPS OF ENGINEERS, North Pacific Division, "User's Manual, SSARR Model, Streamflow Synthesis & Reservoir Regulation," August 1987.

U.S. BUREAU OF RECLAMATION, "Inventory of Hydrologic Models," Global Climate Change Response Program, Denver, Colorado, August 1991.

VIESSMAN, W., G. L. LEWIS, and J. W. KNAPP, *Introduction to Hydrology,* 3rd edition, Harper & Row, 1989.

WANIELISTA, M. P., and Y. A. YOUSEF, *Stormwater Management,* John Wiley & Sons, 1993.

WILLIAMS, J. R., A. D. NICKS, and J. G. ARNOLD, "SWRRB, A Simulator for Water Resources in Rural Basins," *Journal of Hydraulic Engineering,* American Society of Civil Engineers, Vol. 111, No. 6, 1985.

WILLIAMS, J. R., and R. W. HANN, "HYMO: Problem-Oriented Computer Language for Hydrologic Modeling, User's Manual," ARS-S-9, Agricultural Research Service, May 1973.

YOUNG, R. A., C. A. ONSTAD, D. D. BOSCH, and W. P. ANDERSON, "AGNPS, Agricultural Non-Point Source Pollution Model, a Watershed Analysis Tool," Conservation Research Report 35, Agricultural Research Service, U.S. Department of Agriculture, Washington, D.C., 1987.

YOUNG, R. A., C. A. ONSTAD, D. D. BOSCH, and W. P. ANDERSON, "AGNPS, Agricultural Non-Point Source Pollution Model for Evaluating Agricultural Watersheds," *Journal of Soil and Water Conservation,* Vol. 44, No. 2, 1989.

8

STREAM HYDRAULICS MODELS

INTRODUCTION

The computer programs addressed in this chapter simulate flow conditions in natural and improved streams and rivers and associated flood plains and in human-made channels. Some of the programs can also be applied to reservoirs, estuaries, and coastal waters. Model input includes channel geometry and roughness data and either steady-state or time-dependent inflow rates. Steady, varied flow models compute flow depths as a function of location along the channel. Unsteady flow models calculate discharges and flow depths as a function of time and location. Some programs model movable as well as fixed channel beds. Movable boundary models simulate erosion and sediment transport processes as well as flow.

Model Applications

Open channel hydraulics models are typically used in combination with the watershed, river and reservoir water quality, and reservoir/river system operations models described in Chapters 7, 9, and 10. In many cases, hydraulic or hydrodynamic models are an integral component of the models discussed in those chapters.

Flow rates and velocities computed with hydraulic models provide basic input required by water quality models. Velocities are also required for erosion and scour studies.Water surface profiles are needed for many water management applications.

Water supply diversion intake structures may be inoperative if river stages drop below certain levels. Navigation operations and design studies are based on maintaining specified flow depths. Flood-plain management programs require flood-plain delineations based on water surface profiles for floods of specified exceedence frequency. Flood control structures and channel improvements are sized based on design water surface profiles. Reservoir operations are based on river stages. Erosion and sedimentation may also be a significant consideration in design and operation of river control structures.

River Hydraulics Literature

The fundamentals of open channel hydraulics are covered in textbooks by Chow (1959), Henderson (1966), French (1985), and Chaudhry (1993). The U.S. Army Corps of Engineers (1993) provides a comprehensive treatment of river hydraulics. Abbott (1979) and Cunge et al. (1980) focus on numerical solution techniques for the equations describing flow in rivers. The aforementioned textbooks, Davidian (1984), and many other references outline methods for computing water surface profiles for steady, gradually varied flow. Wurbs (1985) provides an annotated bibliography on unsteady flow modeling, with a particular focus on flood waves caused by breached dams. Mahmood and Yevjevich (1975) and Miller and Yevjevich (1975) provide a comprehensive treatment of modeling unsteady flow in open channels. Fread (1993) provides a concise overview of unsteady flow modeling in rivers. The American Society of Civil Engineers (1975), Simons and Senturk (1976), and Petersen (1986) address sediment transport in riverine systems.

Model Categorization

Models are categorized based on capabilities for simulating various types of flow conditions. Although the real world is three-dimensional, flow in rivers can be realistically assumed to be either one- or two-dimensional. Flow can also be characterized as steady versus unsteady and as uniform versus nonuniform or varied. Models can also be categorized as fixed versus movable boundary.

In one-dimensional flow models, accelerations in any direction other than the longitudinal direction of flow are assumed to be negligible. The water surface is assumed to be level perpendicular to the direction of flow. River flows that are significantly influenced by abrupt contractions and expansions in channel or flood-plain topography, bridge embankments, control structures, and/or tributary confluences may be modeled more realistically with equations that describe the water motion in two horizontal directions. The basic flow equations can also be expressed in various forms which reflect components of flow in the three directions of a Cartesian coordinate system. Two-dimensional flow modeling is much more common than three-dimensional modeling. The few fully three-dimensional models that currently exist are in a developmental stage.

Flow is also categorized based on whether flow characteristics (discharge, depth, velocity, etc.) change with location and/or time. Steady versus unsteady refers to whether flow characteristics change with time. Unsteady flows vary with time at a given location. Steady flows are constant over time. Uniform versus nonuniform (varied) refers to whether flow characteristics change with location along the length of the channel. Uniform flow means that discharge, depth, and velocity are the same at each cross-section of the channel reach. Because flow is essentially never simultaneously unsteady and uniform, unsteady flow implies nonuniform flow as well. However, modeling applications are common for steady uniform and both steady and unsteady nonuniform (varied) flow.

Steady uniform flow computations are performed with a uniform flow formula, typically the Manning equation. Steady, gradually varied (nonuniform), one-dimensional water surface profile computations are based on the energy equation. The iterative standard step method solution of the energy equation is commonly applied to streams and rivers and can also be applied to prismatic, human-made channels. The direct step method solution of the energy equation is often applied to prismatic channels but is not applicable to the irregular geometry of rivers. The uniform flow Manning equation is used to approximate energy losses in the nonuniform flow computations.

Two alternative general approaches to one-dimensional unsteady flow modeling are commonly applied. The two alternative approaches are often referred to as hydrologic routing versus hydraulic routing.

Hydraulic routing is based on the St. Venant equations or simplifications of them. The St. Venant equations are two partial differential equations, typically solved numerically, which represent the principles of conservation of mass and momentum. Dynamic routing refers to solution of the complete St. Venant equations. Other hydraulic routing methods, such as diffusion and kinematic routing, are based on simplifications involving neglecting certain terms in the momentum equation. Unlike the hydrologic routing approach, hydraulic routing models simultaneously solve for flow rates (or velocities) and depths (or surface elevations) in the same computational algorithms.

Hydrologic storage routing methods, such as Muskingum, modified Puls, working R and D, and variable storage coefficient, are incorporated in several of the watershed models described in Chapter 7. Hydrologic routing techniques predict the hydrograph (flow rate versus time) at a downstream location, given the hydrograph at an upstream location. Hydrologic routing is used to develop hydrographs at pertinent locations in the stream system. Because hydrologic routing involves only flow rates, if the corresponding flow depths and velocities are needed, the routing results must be combined with a hydraulic model. Typically, the depth and velocity is computed only for the hydrograph peaks (maximum discharges). Use of the Manning equation, developed for steady uniform flow, is the simplest and most approximate approach for associating a flow depth and velocity with the peak discharge from the routing model. A somewhat more accurate approach, which incorporates backwater effects, is to use a steady, gradually varied flow model (such as the standard step method solution of energy equation) to compute the longitudinal water surface profile for the given peak discharges at the various cross-sections.

Addition of erosion and sedimentation considerations makes a model significantly more complex. In mobile bed models, the bed geometry and roughness coefficients are interrelated with the flow hydraulics. Movable boundary models incorporate sediment transport, bed roughness, bed armor, bed surface thickness, bed material sorting, bed porosity, and bed compaction equations as well as the sediment continuity equation, which defines the sediment exchange rate between the water column and bed surface.

The following review addresses four categories of river hydraulics models:

- One-dimensional, steady-state, gradually varied water surface profile models
- One-dimensional unsteady flow models
- Two-dimensional steady and unsteady flow models
- Movable bed models

REVIEW OF AVAILABLE MODELS

One-Dimensional, Steady-State, Gradually Varied Water Surface Profile Models

HEC-2.   The HEC-2 Water Surface Profiles program is an accepted standard for developing water surface profiles. The model was originally developed in the 1960s and has evolved through numerous modifications and expansions. A users manual (USACE Hydrologic Engineering Center 1991), training documents covering specific model capabilities, and papers and reports on specific applications are available from the Hydrologic Engineering Center. Guidelines for applying HEC-2 are also outlined in several books, including Hoggan (1989) and Bedient and Huber (1992). Several universities, as well as the Hydrologic Engineering Center, regularly offer continuing education short courses on applying HEC-2.

Although HEC-2 is a stand-alone model which has been applied in many different modeling situations, it is often used in combination with HEC-1. The HEC-1 Flood Hydrograph Package is discussed in Chapter 7. A typical HEC-1/HEC-2 application involves predicting the water surface profiles which would result from actual or hypothetical precipitation events. Precipitation associated with an actual storm, design storm of specified exceedence frequency, or design storm such as the probable maximum storm is provided as input to the HEC-1 watershed model. HEC-1 performs the precipitation-runoff and routing computations required to develop hydrographs at pertinent locations in the stream system. Peak discharges from the HEC-1 hydrographs are provided as input to HEC-2, which computes the corresponding water surface elevations at specified locations. HEC-2 is also sometimes used to develop discharge versus storage volume relationships for stream reaches which are used in HEC-1 for the modified Puls routing option.

HEC-2 is intended for computing water surface profiles for steady, gradually varied flow in natural or human-made channels. The computational procedure is based on the solution of the one-dimensional energy equation, with energy losses due to friction estimated with the Manning equation. This computational procedure is generally known as the standard step method. The computations proceed by reach, with known values at one cross-section used to compute the water surface elevation, mean velocity, and other flow characteristics at the next cross-section. Either subcritical or supercritical flow regimes can be modeled. The effects of obstructions to flow such as bridges, culverts, weirs, and buildings located in the flood plain may be reflected in the model.

HEC-2 provides a number of optional capabilities related to bridges and culverts, channel improvements, encroachments, tributary streams, split streams, ice-covered streams, effective flow areas, interpolated cross-sections, friction loss equations, critical depth, multiple profiles, developing storage-outflow data, and calibrating Manning roughness coefficients.

WSPRO. WSPRO is a water surface profile model developed by the U.S. Geological Survey for the Federal Highway Administration (Shearman 1990). WSPRO is similar to HEC-2 and is also based on the standard step method solution of the energy equation. The program is designed for modeling natural rivers with irregular geometry. One-dimensional, gradually varied flow is assumed. Both subcritical and supercritical flow may be modeled. The motivation for developing WSPRO was to expand capabilities for analyzing the impacts of bridges on water surface profiles, in response to Federal Highway Administration requirements for thorough evaluation of design alternatives for the hydraulics of bridge waterways. Comprehensive options are provided for simulating flow through bridges and culverts and over embankments. Multiple culverts at a road crossing may be analyzed.

Other water surface profile models. Eichert (1970) surveyed generalized water surface profile computer programs available in 1969; compiled information on 11 models; and compared HEC-2 and five other selected models developed by the USACE Little Rock District, Bureau of Reclamation, Soil Conservation Service, U.S. Geological Survey, and Iowa Natural Resources Council. Several of these models continue to be used, and other water surface profile models have been developed since then. The Soil Conservation Service (1976) WSP2 computer program has been particularly widely used for many years.

The A&M Watershed Model, discussed in Chapter 7, includes computation of water surface profiles using the standard step method solution of the energy equation. Whereas HEC-1 and HEC-2 are separate stand-alone programs which are often used in combination, the A&M Watershed Model combines precipitation-runoff, routing, and water surface profile computations in the same model.

HEC-RAS. The USACE Hydrologic Engineering Center is in the process of developing a next generation (NexGen) of hydrologic engineering software. The River Analysis System (HEC-RAS) package, developed in conjunction with the NexGen project, will provide capabilities for simulating one-dimensional steady or unsteady flow and also sediment transport and movable boundary open channel flow. The HEC-RAS system will ultimately contain three components for (1) steady flow water surface profile computations; (2) unsteady flow computations; and (3) movable boundary hydraulic computations. All three components will use a common geometric data representation, as well as common geometric and hydraulic computation routines. HEC-RAS will be an integrated system, designed for use in a multitasking, multiuser environment. The system will be comprised of a graphical user interface, separate computational engines, data storage/management components, graphics, and reporting capabilities. An initial version of the HEC-RAS focusing on steady flow water surface profile computations, as of 1994, is undergoing testing prior to public release.

One-Dimensional Unsteady Flow Models

The aforementioned steady-state, gradually varied water surface profile models are based on the classic energy equation, which does not include time. As previously discussed, unsteady (time varying) flow problems are often modeled by combining hydrologic routing with a steady-state hydraulic model. HEC-2-type steady-state models are applied to unsteady flow problems based on the assumption that approximately steady-state conditions occur temporarily as the hydrographs peak. Several watershed models cited in Chapter 7 (including HEC-1, TR-20, A&M Watershed Model, and SSARR) include hydrologic flood routing options. Hydrologic routing is an approximate unsteady flow modeling approach which does not capture backwater and acceleration effects. Dynamic routing is required to represent more accurately backwater and acceleration effects in unsteady flow modeling.

Dynamic routing is based on solving the complete St. Venant equations representing conservation of mass and momentum. Other unsteady flow hydraulic modeling techniques are based on neglecting certain terms or otherwise simplifying the St. Venant equations. These various simplified hydraulic routing methods are conceptually similar to various variations of hydrologic routing combined with steady-state hydraulic models.

Most dynamic routing models have been developed for specific applications and are not generalized operational models. The Tennessee Valley Authority developed one of the earlier generalized dynamic routing models, which was subsequently modified by the USACE Waterways Experiment Station and USACE Hydrologic Engineering Center. National and state dam safety programs provided a major impetus for development of operational generalized dynamic routing programs during the late 1970s and early 1980s (Wurbs 1985).

Several of the water quality modeling packages discussed in Chapter 9 contain unsteady flow models. The Water Quality for River-Reservoir Systems (WQRRS) model includes the Stream Hydraulics Package (SHP), which includes options for steady-state water surface profile computations; Muskingum and modified Puls hydrologic routing; kinematic routing; and a finite element solution of the St. Venant equations (dynamic routing). HEC-5 and HEC-5Q contain several hydrologic routing options. The dynamic routing component of the CE-QUAL-RIV1 model is based on an implicit finite difference solution of the St. Venant equations. The Water Quality Analysis Simulation Program (WASP) includes the unsteady flow hydrodynamic program DYNHYD, which is based on a simplified hydraulic routing approach. The Stormwater Management Model (SWWM) discussed in Chapter 7 includes the stand-alone dynamic routing model EXTRAN.

NWS models. The Operational Dynamic Wave Model (DWOPER), Dam-Break Flood Forecasting Model (DAMBRK), and Flood Wave (FLDWAV) are dynamic routing models developed by the Hydrologic Research Laboratory of the National Weather Service. DAMBRK is a specific-purpose dam-breach model that stemmed from the general-purpose DWOPER. The more recent FLDWAV combines the capabilities of DWOPER and DAMBRK. FLDWAV was developed with the intention of replacing DWOPER and DAMBRK, but all three models continue to be used widely. FLDWAV is included in the Model Inventory Appendix. The one-dimensional unsteady flow models are based on an implicit finite difference solution of the complete St. Venant equations. Discharges, velocities, depths, and water surface elevations are computed as a function of time and distance along the river (Fread 1987, 1988; Fread and Lewis 1988). The models have been used on IBM-compatible microcomputers as well as minicomputers and mainframes.

DWOPER is used routinely by the National Weather Service River Forecast Centers and has also been applied widely outside the National Weather Service. DWOPER has wide applicability to rivers of varying physical features, such as branching tributaries, irregular geometry, variable roughness parameters, lateral inflows, flow diversions, off-channel storage, local head losses such as bridge contractions and expansions, lock and dam operations, and wind effects. An automatic calibration feature is provided for determining values for roughness coefficients. Data management features facilitate use of the model in a day-to-day forecasting environment. The model is equally applicable for simulating unsteady flows in planning and design studies.

DAMBRK has been applied extensively by various agencies and consulting firms in conducting dam safety studies. DAMBRK simulates the failure of a dam, computes the resultant outflow hydrograph, and simulates the movement of the flood wave through the downstream river valley. An inflow hydrograph is routed through a reservoir optionally using either hydrologic storage routing or dynamic routing. Two types of breaching may be simulated. An overtopping failure is simulated as a rectangular, triangular, or trapezoidal opening that grows progressively downward from the dam crest with time. A piping failure is simulated as a rectangular orifice that grows with time and is centered at any specified elevation within the dam. The pool elevation at which breaching begins, time required for breach formation, and geometric parameters of the breach must be specified by the user. The DWOPER dynamic routing algorithm is used to route the outflow hydrograph through the downstream valley. DAMBRK can simulate flows through multiple dams located in series on the same stream.

DWOPER does not include the dam-breach modeling capabilities of DAMBRK. DAMBRK is limited to a single river without tributaries and thus does not provide the flexibility of DWOPER in simulating branching tributary configurations.

FLDWAV combines DWOPER and DAMBRK into a single model and provides additional hydraulic simulation methods within a more user-friendly model structure (Fread and Lewis 1988). FLDWAV, like DWOPER and DAMBRK, is based on an expanded form of the St. Venant equations that includes the following hydraulic effects: lateral inflows and outflows; off-channel storage; expansion and contraction losses; mixed subcritical and supercritical flow; nonuniform velocity distribution across the flow section; flow path differences between the flood plain and a sinuous main channel; and surface wind shear. The model can simulate dam breaches in one or several dams located sequentially on the same stream. Other conditions that can be simulated include levee overtopping; interactions between channel and flood-plain flow; and combined free-surface and pressure flow. FLDWAV also has a calibration option for determining Manning roughness coefficient values.

UNET. UNET simulates one-dimensional unsteady flow through a full network of open channels (Barkau 1993). In addition to solving the network system, UNET provides the user the ability to apply several external and internal boundary conditions, including flow and stage hydrographs, rating curves, gated and uncontrolled spillways, pump stations, bridges, culverts, and levees. The dynamic routing model is based on a four-point linear implicit finite difference solution of the St. Venant equations. UNET is conceptually similar to the National Weather Service DWOPER and FLDWAV models described earlier. UNET interconnects with the Hydrologic Engineering Center Data Storage System (HEC-DSS) described in Chapter 3. The UNET system also allows use of HEC-2-style cross-section input data files.

UNET is a proprietary program originally developed by Dr. R. L. Barkau. An earlier unsteady flow, open channel network program called USTDY was developed by Dr. Barkau while previously employed by the Corps of Engineers. Under an agreement with the developer, the Hydrologic Engineering Center maintains a version of UNET with documentation for distribution and use both within and outside the Corps of Engineers.

Two-Dimensional Steady and Unsteady Flow Models

In two-dimensional river modeling, acceleration in the vertical direction is typically assumed to be negligible, with the equations being formulated for the two horizontal dimensions. Two-dimensional modeling is much more complex and requires much more input data to describe the channel geometry and flow resistance characteristics than one-dimensional modeling. Lee and Froehlich (1986) review the literature on two-dimensional river modeling and provide an extensive list of references.

RMA-2 and FESWMS-2DH are the two-dimensional riverine system flow models which are probably most widely used in the United States. RMA-2 and FESWMS-2DH are similar. Both are fully implicit finite element models that use the primitive formulation of the fully nonlinear shallow water equations. Both include options for modeling either steady or unsteady flows. RMA-2 was developed by Resource Management Associates under contract with the USACE (Norton et al. 1973). The model has undergone extensive revisions in the 20 years since it was originally developed. Thomas et al. (1990) document a recent version of RMA-2. Versions of RMA-2, along with associated pre- and postprocessor software, are available from the USACE Waterways Experiment Station and Hydrologic Engineering Center. FESWMS-2DH is discussed later in this chapter and in the Model Inventory Appendix.

The Diffusion Hydrodynamic Model (DHM) is a generalized, two-dimensional, unsteady flow model developed by the U.S. Geological Survey (Hromadka and Yen 1987). The DHM was originally developed for a two-dimensional dam-breach study and subsequently expanded for other applications. DHM is based on the noninertial form of the St. Venant equations for two-dimensional flow. An option allows for simpler kinematic wave calculations instead of the diffusion routing.

FESWMS-2DH. The Finite Element Surface-Water Modeling System: Two-Dimensional Flow in a Horizontal Plane (FESWMS-2DH) was developed by the U.S. Geological Survey, Water Resources Division, for the Federal Highway Administration (Froehlich 1989). The primary motivation for developing FESWMS-2DH was to improve capabilities for analyzing flow at highway bridge crossings, where complicated hydraulic conditions exist. However, the generalized model is applicable to a broad range of different steady and unsteady two-dimensional river modeling situations.

FESWMS-2DH is a modular set of computer programs which includes DINMOD, the data input module; FLOMOD, the depth-averaged flow solution module; and ANOMOD, the analysis of output module. DINMOD prepares a finite element network for use by other FESWMS-2DH programs. ANOMOD generates plots and reports of computed values that simplify interpretation of simulation results. FLOMOD solves the vertically integrated conservation of momentum equations and the conservation of mass (continuity) equation, using the Galerkin finite element method, to obtain depth-averaged velocities and water depth at points in a finite element network. Energy losses are computed using the Chezy or Manning equations and Boussinesq eddy viscosity concept. The effects of wind stress and the Coriolis force may be included in the simulation. The model is capable of simulating flow through single or multiple bridge openings as normal flow, pressure flow, weir flow, or culvert flow.

Movable Boundary Models

In movable boundary or mobile bed models, the channel geometry and roughness characteristics change during the simulation due to erosion and sedimentation processes. Flow hydraulics and sediment transport are interrelated.

HEC-6. The HEC-6 Scour and Deposition in Rivers and Reservoirs program (USACE Hydrologic Engineering Center 1993) is designed to simulate long-term trends of scour and/or deposition in a stream channel resulting from modifying flow frequency and duration or channel geometry. HEC-6 can be used to predict reservoir sedimentation; design channel contractions required to maintain navigation depths or decrease the volume of maintenance dredging; analyze the impacts of dredging on deposition rates; evaluate sedimentation in fixed channels; and estimate maximum scour during floods. HEC-6 features include capabilities for simulating stream networks, channel dredging, and levee and encroachment alternatives.

HEC-6 is a one-dimensional, movable boundary numerical model designed to simulate and predict changes in river profiles resulting from scour and/or deposition over moderate time periods. Although single flood events can be analyzed, the model is oriented toward evaluating trends over a number of years. A continuous flow record is partitioned into a series of steady flows of variable duration. Water surface profile computations are performed for each flow. Potential sediment transport rates are then computed at each section. These rates, combined with the duration of the flow, permit a volumetric accounting of sediment within each reach. The amount of scour or deposition at each section is then computed and the cross-section adjusted accordingly. The computations then proceed to the next flow in the sequence, and the cycle is repeated beginning with the updated channel geometry. The sediment calculations are performed by grain size fraction, thereby allowing the simulation of hydraulic sorting and armoring.

The HEC-6 water surface profile computations are based on a standard step method solution of the one-dimensional, steady-state energy equation, as for HEC-2. HEC-6 simulates the capability of a stream to transport sediment given the yield from upstream sources. Sediment transport includes both bed and suspended load. The computational formulation utilizes Einstein's basic concepts of sediment transport. A number of user-selected alternative transport functions for bed material load are included in the model.

TABS-2. The Open Channel Flow and Sedimentation (TABS-2) system was developed and is maintained by the USACE Waterways Experiment Station (Thomas and McAnally 1985). TABS-2 is a collection of generalized simulation models and utility programs integrated into a numerical modeling system for studying two-dimensional hydraulics, transport, and sedimentation processes in rivers, bays, and estuaries. Major components of the system include RMA-2V and STUDH for simulating hydrodynamics and sediment transport, respectively, and graphics display programs. TABS-2 computes water surface elevations, current patterns, dispersive transport, sediment erosion, transport and deposition, resulting bed surface elevations, and feedback to hydraulics. Existing and proposed geometry can be analyzed to determine the impact of project designs on flows, sedimentation, and salinity. The two-dimensional model is particularly useful for analyzing velocity patterns around structures and islands.

COMPARISON OF MODELS INCLUDED IN THE MODEL INVENTORY APPENDIX

The following models are included in the Model Inventory Appendix:

HEC-2: Water Surface Profiles Model—USACE Hydrologic Engineering Center

WSPRO: Water Surface Profiles Model—U.S. Geological Survey and Federal Highway Administration

FLDWAV: Flood Wave Model—National Weather Service

UNET: One-Dimensional Unsteady Flow through a Full Network of Open Channels—USACE Hydrologic Engineering Center

FESWMS-2DH: Finite Element Surface-Water Modeling System: Two-Dimensional Flow in a Horizontal Plane—U.S. Geological Survey and Federal Highway Administration

HEC-6: Scour and Deposition in Rivers and Reservoirs—USACE Hydrologic Engineering Center.

The six alternative software packages are reasonably well documented and include executable versions available for MS-DOS-based microcomputers. The FORTRAN77 programs have been compiled and executed on other computer systems as well. HEC-2 and WSPRO are the simplest of the six models. HEC-2 is the most widely used. The other more complex models are used whenever the simplifying assumptions inherent in HEC-2 (or the combined HEC-1 and HEC-2) are inappropriate for a particular application. The four models can be compared based on their inherent assumptions regarding hydraulic flow conditions. HEC-2 and WSPRO are one-dimensional, steady, gradually varied flow models. FLDWAV and UNET are one-dimensional unsteady flow models. FESWMS-2DH is a two-dimensional model which can be used optionally for simulating either steady or unsteady flows. The one-dimensional, movable bed HEC-6 combines sediment transport with steady flow computations.

The simpler HEC-2 and WSPRO compute water surface profiles (elevations along the stream) for inputted discharges at specified cross-sections. The flow is assumed to be one-dimensional, with the water surface being level perpendicular to the longitudinal flow direction. Time is not reflected in HEC-2. A common approach for applying HEC-2 involves inputting peak discharges, which have been computed by routing hydrographs through stream reaches using hydrologic storage routing methods incorporated in HEC-1 or a similar model. This approach is reasonably valid as long as flow characteristics change relatively slowly over time and the discharges are not subject to backwater effects. As acceleration effects or changes in flow characteristics over time become more important, a dynamic wave model such as FLDWAV or UNET is used. Dynamic routing also incorporates backwater effects into the flow computations. Dynamic routing couples the determination of discharges and water surface elevations as a function of both time and location. FESWMS-2DH is used in situations in which two-dimensional flow effects are important. Such applications could involve flow contractions and expansions caused by bridge abutments, abrupt changes in topography, or stream tributary confluences.

REFERENCES

ABBOTT, M. B., *Computational Hydraulics, Elements of the Theory of Free Surface Flows,* Pitman Publishers, London, England, 1979.

AMERICAN SOCIETY OF CIVIL ENGINEERS, "Sedimentation Engineering," Manual 54, V. A. Vanoni, editor, New York, 1975.

BARKAU, R. L., "UNET, One-Dimensional Flow through a Full Network of Open Channels, User's Manual, Version 2.1," CPD-66, U.S. Army Corps of Engineers, Hydrologic Engineering Center, Davis, California, May 1993.

BEDIENT, P. B., and W. C. HUBER, *Hydrology and Floodplain Analysis,* 2nd edition, Addison-Wesley, 1992.

CHAUDHRY, M. H., *Open-Channel Flow,* Prentice Hall, 1993.

CHOW, V. T., *Open-Channel Hydraulics,* McGraw-Hill, 1959.

CUNGE, J. A., F. M. Holley, and A. Verway, *Practical Aspects of Computational River Hydraulics,* Pitman Publishers, London, England, 1980.

DAVIDIAN, Jacob, "Computation of Water-Surface Profiles in Open Channels," Book 3, Chapter A15, Techniques of Water-Resources Investigations of the U.S. Geological Survey, USGS, Alexandria, Virginia, 1984.

EICHERT, B. S., "Survey of Programs for Water-Surface Profiles," *Journal of the Hydraulics Division,* American Society of Civil Engineers, Vol. 96, No. HY2, February 1970.

FREAD, D. L., "Flood Routing," in *Handbook of Hydrology,* edited by D. R. Maidment, McGraw-Hill, 1993.

FREAD, D. L., "National Weather Service Operational Dynamic Wave Model," Hydrologic Research Laboratory, National Weather Service, Silver Spring, Maryland, April 1978, reprinted April 1987.

FREAD, D. L., "The NWS DAMBRK Model, Theoretical Background and User Documentation," HDR-256, Hydrologic Research Laboratory, National Weather Service, Silver Spring, Maryland, June 1988.

FREAD, D. L., and J. M. Lewis, "FLDWAV: A Generalized Flood Routing Model," Proceedings of the 1988 National Conference on Hydraulic Engineering (S. R. Abt and J. Gessler, editors), American Society of Civil Engineers, 1988.

FRENCH, R. H., *Open-Channel Hydraulics,* McGraw-Hill, 1985.

FROEHLICH, D. C., "Finite Element Surface-Water Modeling System: Two-Dimensional Flow in a Horizontal Plane, User's Manual," FHWA-RD-88-177, Federal Highway Administration, Turner-Fairbank Highway Research Center, McLean, Virginia, April 1989.

HENDERSON, F. M., *Open-Channel Hydraulics,* Macmillan, 1966.

HOGGAN, D. H., *Computer-Assisted Hydrology & Hydraulics Featuring the U.S. Army Corps of Engineers HEC-1 and HEC-2 Software System,* McGraw-Hill, 1989.

HROMADKA, T. V., and C. C. YEN, "A Diffusion Hydrodynamic Model," U.S. Geological Survey Water Resources Investigations Report 87-4137, U.S. Government Printing Office, Washington, D.C., 1987.

LEE, J. K., and D. C. FROEHLICH, "Review of Literature on the Finite-Element Solution of the Equations of Two-Dimensional Surface-Water Flow in the Horizontal Plane," Circular 1009, U.S. Geological Survey, Denver, Colorado, 1986.

MAHMOOD, K., and V. YEVJEVICH (editors), *Unsteady Flow in Open Channels,* volumes I and II, Water Resources Publications, Littleton, Colorado, 1975.

MILLER, W. A., and V. YEVJEVICH, *Unsteady Flow in Open Channels, Volume III, Bibliography,* Water Resources Publications, Littleton, Colorado, 1975.

NORTON, W. R., I. P. KING, and G. T. ORLOB, "A Finite Element Model for Lower Granite Reservoir, Vol. 3, Appendix F," Water Quality Report, Lower Granite Lock and Dam, Snake River, Washington-Idaho, U.S. Army Corps of Engineers, Walla Walla District, Walla Walla, Washington, 1973.

PETERSEN, M. S., *River Engineering,* Prentice Hall, 1986.

SHEARMAN, J. O., "User's Manual for WSPRO, a Computer Model for Water Surface Profile Computation," Report No. FHWA-IP-89-027, U.S. Geological Survey, 1990.

SIMONS, D. B., and F. SENTURK, *Sediment Transport Technology,* Water Resources Publications, Fort Collins, Colorado, 1976.

SOIL CONSERVATION SERVICE, "WSP2 Computer Program, Technical Release 61," Engineering Division, USDA, 1976.

THOMAS, W. A., and W. H. MCANALLY, "Open-Channel Flow and Sedimentation TABS-2, Users's Manual," Instruction Report HL-85-1, U.S. Army Engineer Waterways Experiment Station, Vicksburg, Mississippi, 1985.

THOMAS, W. A., W. H. MCANALLY, and J. V. LETTER, "Appendix F: User Instructions for RMA-2V, a Two-Dimensional Model for Free-Surface Flows," Draft, Generalized Computer Program System for Open-Channel Flow and Sedimentation, U.S. Army Engineer Waterways Experiment Station, Vicksburg, Mississippi, 1990.

U.S. ARMY CORPS OF ENGINEERS, "River Hydraulics," Engineering Manual 1110-2-1416, Washington, D.C., October 1993.

U.S. ARMY CORPS OF ENGINEERS, Hydrologic Engineering Center, "HEC-2 Water Surface Profiles, User's Manual," Davis, California, September 1990 (revised February 1991).

U.S. ARMY CORPS OF ENGINEERS, Hydrologic Engineering Center, "HEC-6 Scour and Deposition in Rivers and Reservoirs, User's Manual," Davis, California, August 1993.

WURBS, R. A., "Military Hydrology Report 9: State-of-the-Art Review and Annotated Bibliography of Dam-Breach Flood Forecasting," MP EL-79-6, U.S. Army Engineer Waterways Experiment Station, Vicksburg, Mississippi, February 1985.

9

RIVER AND RESERVOIR WATER
QUALITY MODELS

INTRODUCTION

This chapter focuses on models for simulating water quality conditions in streams, rivers, lakes, and reservoirs. Some of the models cited may also be applied to estuaries and coastal waters. Precipitation-runoff models, which include predicting the water quality of watershed runoff, were covered in Chapter 7. Ground-water quality was addressed in Chapter 6.

Model Applications

Models provide a means to predict the impacts of natural processes and human activities on the physical, chemical, and biological characteristics of water in a river/reservoir system. Models are used widely to evaluate the impacts of waste loads from treatment plants or pollutant loads from various other point and nonpoint sources. Alternative reservoir operating plans can be evaluated from the perspective of the effects of releases on in-pool and downstream water quality. Models can be used in conjunction with water quality monitoring activities to interpolate or extrapolate sampled data to other locations and times. Models are also used as research tools to develop an understanding of the processes and interactions affecting water quality.

Hydrodynamic and Water Quality Parameters

Water quality models simulate hydrodynamic as well as water quality conditions. Hydrodynamics and water quality are often reflected in separate interfacing submodels. The models incorporate equations for transport or conservation of mass for water and equations for transport and transformation of materials in the water. The hydrodynamic equations deal with water volumes, flow rates, velocities, and depths. The materials transport and transformation equations include expressions for energy transfer and expressions for chemical equilibrium or chemical and biological kinetics. Classical formulations for nutrient uptake, growth, photosynthesis, predation, and microbial decomposition have been used in many models.

Water quality constituents can be categorized as organic, inorganic, radiological, thermal, and biological. Pollutants may be classified by specific forms, such as biochemical oxygen demand, nitrogen, phosphorus, bacteria, or specific toxic substances. Unstable pollutants, which decay with time, are termed nonconservative. Many inorganic pollutants are treated as being conservative. Pollutants may be loaded into a watercourse from either point or nonpoint sources. Water quality parameters that have been modeled include temperature, total dissolved solids or inorganic salts, inorganic suspended sediments, dissolved oxygen, biochemical nutrients (phosphorus, nitrogen, silicon), inorganic carbon (carbon, pH), biomass and food chains (chlorophyll α, zooplankton, etc.), metals (lead, mercury, cadmium), synthetic organic chemicals (polychlorinated biphenyls), radioactive materials (radium, plutonium), and herbicides and insecticides (Dieldrin, DDT).

Parameter values for water quality models often must be developed through field and laboratory studies. Initial estimates can sometimes be approximated from information available in the literature. Sets of parameter values are required for model calibration and verification. Parameter values are also required as model input to define boundary and initial conditions.

Boundary conditions typically include specification of flows and loads entering the river/reservoir system. Boundary conditions also include specification of meteorological conditions that govern the calculation of the energy flux through the water surface and chemical or biological characteristics that govern the exchange of material, such as dissolved oxygen, between the bed and water column. Flow models may also require specification of flow or depth at the downstream end of the model domain to take into account backwater effects.

Initial conditions are required for dynamic models to define parameter values at the beginning of the simulation period. Initial conditions are typically defined by initial values of flow, depth, and all water quality parameters included in the model.

Water Quality Modeling Literature

Books covering the fundamentals of water quality modeling include Tchobanoglous and Schroeder (1985) and Thomann and Mueller (1987). Chapra and Reckhow (1983) treat water quality of lakes. McCutcheon (1989) provides an in-depth coverage of transport and surface exchange in rivers. Orlob (1984, 1992) and Stefan et al. (1989) provide state-of-the-art overviews of surface water quality modeling.

Water quality modeling dates back to development of the Streeter and Phelps oxygen sag equation in conjunction with a study of the Ohio River in the 1920s. One of the first computer models of water quality was developed for a study of the Delaware Estuary (Thomann 1963). The Delaware Estuary model extended the Streeter and Phelps equation to a multisegment system. A number of models were developed and applied during the 1960s and early 1970s. During that period, water quality modeling focused on temperature, dissolved oxygen, and biochemical oxygen demand. By the mid-1970s, the need was apparent for development of capabilities for analyzing a more comprehensive range of water quality parameters. The literature of the last 20 years is extensive. A number of models representative of the state of the art are cited in a later section of this chapter.

Model Categorization

Standing bodies of water (lakes and reservoirs) are somewhat different from flowing streams and rivers. A model may include capabilities for simulation of either rivers or reservoirs or both. Capabilities for modeling the vertical distribution of temperature and chemical and biological parameters in reservoirs is a particularly important feature included in some models. Models may be pertinent to estuaries and coastal waters as well as lakes, ponds, reservoirs, streams, and rivers.

Models may be categorized as zero-dimensional, one-dimensional, two-dimensional, or three-dimensional. Although two-dimensional river models are not uncommon, rivers are typically treated as one-dimensional, with values for water quality and flow parameters changing only in the longitudinal direction. Zero-dimensional and multiple-dimensional modeling is more commonly associated with reservoirs. Reservoirs are also often treated as one-dimensional, with gradients in the vertical direction. Thus, one-dimensional models typically reflect longitudinal gradients for rivers and vertical gradients for reservoirs. Zero-dimensional input-output models contain no information on hydrodynamics other than the assumption that the water body is well mixed. Three-dimensional models include vertical, lateral, and longitudinal changes in water quality and flow parameters.

In steady-state models, parameter values do not vary with time. Dynamic or unsteady models allow parameters to vary with time. Steady versus unsteady refers to both flow parameters (discharge, velocity, depth) and water quality parameters.

In a fully dynamic model, both flow and water quality parameters are unsteady. In some models, some parameters are allowed to vary with time whereas other parameters are assumed to be steady state.

For purposes of the following review of available models, water quality models are grouped as follows:

- Steady-state models for riverine systems
- Dynamic models for riverine systems
- One-dimensional (vertical) models for reservoirs and lakes
- Multidimensional models for reservoirs and lakes

REVIEW OF AVAILABLE MODELS

The steady-state and dynamic riverine system water quality models cited in this section provide capabilities for simulating reservoirs as components of the overall aquatic system. Some reflect only one-dimensional longitudinal gradients. However, others such as WASP, WQRRS, and HEC-5 include features for modeling vertical stratification in reservoirs along with a range of other simulation capabilities. However, models which focus specifically on vertical stratification in reservoirs and lakes are grouped and discussed later in this chapter.

Steady-State Models for Riverine Systems

QUAL2E. The Enhanced Stream Water Quality Model (QUAL2E) is the only steady-state riverine quality model included in the Model Inventory Appendix. QUAL2E is actually a basically steady-state model with some dynamic features. Steady-state hydraulics (flows) are combined with water quality parameters that can optionally be steady-state or reflect daily variations.

QUAL2E has been used widely and is an accepted standard, particularly for waste-load allocation studies of stream systems. The current release of the Environmental Protection Agency model (Brown and Barnwell 1987) is included in the Model Inventory Appendix. A number of other versions have been developed and applied by various entities. QUAL2E and its variations stem from early models developed by the Texas Water Development Board (TWDB). The TWDB DOSAG model solves the steady-state oxygen sag problem for a multisegment river reach. QUAL I (TWDB 1971) was developed by expanding DOSAG. QUAL II (Roesner et al. 1973) was developed for the Environmental Protection Agency by expanding and improving QUAL I.

QUAL2E is a one-dimensional (longitudinal) model for simulating well-mixed streams and lakes (Brown and Barnwell 1987). A watercourse is represented as a series of piece-wise segments or reaches of steady nonuniform flow. Flows are constant with time and uniform in each reach but can vary from reach to reach. QUAL2E allows simulation of point and nonpoint loadings, withdrawals, branching tributaries, and in-stream hydraulic structures. The model allows simulation of 15 water quality constituents, including dissolved oxygen, biochemical oxygen demand, temperature, algae as chlorophyll α, organic nitrogen, ammonia nitrogen, nitrate nitrogen, organic phosphorus, inorganic phosphorus, coliforms, an arbitrary nonconservative constituent, and three arbitrary conservative constituents.

QUAL2E has optional features for analyzing the effects on water quality— primarily dissolved oxygen and temperature—caused by diurnal variations in meteorological data. Diurnal dissolved oxygen variations caused by algal growth and respiration can also be modeled. QUAL2E also has an option for determining flow augmentation required to meet any prespecified dissolved oxygen level.

QUAL2E-UNCAS is an enhanced version of QUAL2E which provides capabilities for uncertainty analysis. The uncertainty analysis capabilities include (1) sensitivity analysis with an option for factorially designed combinations of input variable perturbations; (2) first-order error analysis with output consisting of a normalized sensitivity coefficient matrix and a components of variance matrix; and (3) Monte Carlo simulation with summary statistics and frequency distributions of the output variables.

Other steady-state models. Other simpler one-dimensional, steady-state models for analyzing water quality in stream networks include the Waterways Experiment Station STEADY model (Martin 1986) and U.S. Geological Survey Streeter-Phelps model (Bauer et al. 1979). STEADY computes temperature, dissolved oxygen, and biochemical oxygen demand. The USGS model simulates dissolved oxygen, nitrogenous oxygen demand, carbonaceous biochemical oxygen demand, orthophosphate phosphorus, total and fecal coliform bacteria, and three conservative substances.

Two-dimensional, steady-state riverine water quality models have not been applied widely. Examples of two-dimensional models include SARAH (Ambrose and Vandergrift 1986), RIVMIX (Krishnappan and Lau 1985), Ontario Ministry of the Environment models (Gowda 1984), and USGS model (Bauer and Yotsukuru 1974).

Dynamic Models for Riverine Systems

With unsteady or dynamic models, the flow and water quality variables vary as a function of time. The WASP, CE-QUAL-RIV1, WQRRS, and HEC-5 models, included in the Model Inventory Appendix, are dynamic models for simulating river/reservoir systems.

WASP. The Water Quality Analysis Simulation Program (WASP), maintained by the Environmental Protection Agency, is a generalized modeling framework for simulating aquatic systems, including rivers, reservoirs, estuaries, and coastal waters. The various versions of the model evolved from the original WASP (DiToro et al. 1983), which incorporated concepts from a number of other earlier models. As of 1994, Version 4 WASP4 (Ambrose et al. 1988) is being replaced by the newer version WASP5 (Ambrose et al. 1993).

WASP is designed to provide a flexible modeling system. The time varying processes of advection, dispersion, point and diffuse mass loading, and boundary exchange are represented in the basic program. Water quality processes are modeled in special kinetic subroutines that are either selected from a library or supplied by the user. WASP is structured to permit easy substitution of kinetic subroutines into an overall package to form problem-specific models. A compartment modeling approach represents the aqueous system as segments which can be arranged in one, two, or three dimensions.

WASP consists of two stand-alone computer programs, DYNHYD and WASP, that can be run in conjunction or separately. The hydrodynamics program DYNHYD simulates the movement of water. The water quality program WASP models the movement and interaction of pollutants within the water. EUTRO and TOXI are submodels which can be incorporated into the water quality program. EUTRO is used to analyze conventional pollution involving dissolved oxygen, biochemical oxygen demand, nutrients, and eutrophication. TOXI simulates toxic pollution involving organic chemicals, metals, and sediment.

CE-QUAL-RIV1. The Waterways Experiment Station model CE-QUAL-RIV1 is a fully dynamic one-dimensional flow and water quality simulation model for streams (USACE Waterways Experiment Station 1990). The original version of the model was developed at Ohio State University for the Environmental Protection Agency (Bedford et al. 1983), primarily for predicting water quality associated with stormwater runoff. The present CE-QUAL-RIV1 has been modified to handle control structures. The model is designed for analyzing highly unsteady streamflow conditions, such as that associated with peaking hydropower tailwaters. The model also allows simulation of branched river systems with multiple control structures, such as reregulation dams and navigation locks and dams.

The CE-QUAL-RIV1 package includes two stand-alone programs, RIV1H and RIV1Q, which can be interfaced or used separately. RIV1H performs hydraulic routing based on a numerical solution of the full St. Venant equations. RIV1Q is the water quality program. The model is similar to QUAL2E in that it simulates temperature, dissolved oxygen, biochemical oxygen demand, and nutrient kinetics. However, the dynamic CE-QUAL-RIV1 can model sharp flow and water quality gradients.

WQRRS. The Water Quality for River-Reservoir Systems (WQRRS) is a package of dynamic water quality and hydrodynamic models (USACE Hydrologic Engineering Center 1985). The WQRRS package includes the models SHP, WQRRSQ, and WQRRSR, which interface with each other. The Stream Hydraulics Package (SHP) and Stream Water Quality (WQRRSQ) programs simulate flow and quality conditions for stream networks, which can include branching channels and islands. The Reservoir Water Quality (WQRRSR) program is a one-dimensional model used to evaluate the vertical stratification of physical, chemical, and biological parameters in a reservoir. The SHP provides a range of optional methods for computing discharges, velocities, and depths as a function of time and location in a stream system. The hydraulic computations can be performed optionally using input stage-discharge relationships, hydrologic routing, kinematic routing, steady flow equations, or the full unsteady flow St. Venant equations. The WQRRSR and WQRRSQ programs provide capabilities for analyzing up to 18 constituents, including chemical and physical constituents (dissolved oxygen, total dissolved solids), nutrients (phosphate, ammonia, nitrite, and nitrate), carbon budget (alkalinity, total carbon), biological constituents (two types of phytoplankton, benthic algae, zooplankton, benthic animals, three types of fish), organic constituents (detritus, organic sediment), and coliform bacteria.

HEC-5Q. The HEC-5 Simulation of Flood Control and Conservation Systems model is described in Chapter 10. The quantity-only HEC-5 is widely used. A water quality version HEC-5Q (USACE Hydrologic Engineering Center 1986) is also available but has not been used as widely. The HEC-5 model discussed in Chapter 10 provides the flow simulation module for the water quality version HEC-5Q. Additional subroutines provide the water quality simulation module. The water quality module accepts system flows generated by the flow module and computes the vertical distribution of temperature and other constituents in the reservoirs and the water quality in the associated downstream reaches. The water quality module also includes an option for selecting the gate openings for reservoir selective withdrawal structures to meet user-specified water quality objectives at downstream control points.

The water quality model can be applied in three alternative modes: calibration, annual simulation, and long-term mode. In the calibration mode, values of parameters such as decay rates and dispersion coefficients are computed based on inputted historical flow, water quality, and reservoir operation data. In the annual simulation mode, the model uses a daily computational time interval in determining the effects of reservoir operations on the water quality in the reservoirs and downstream river reaches. The long-term mode is similar to the annual mode except that the time steps are longer (generally 30 days), so the effects of reservoir operations on water quality can be examined over longer planning horizons of several or many years.

Water quality constituents vary with two alternative simulation options. With the first option, the following constituents can be included in the model: water temperature (always required); up to three conservative constituents; up to three nonconservative constituents; and dissolved oxygen. The other option, referred to as the phytoplankton option, requires the following eight constituents: water temperature, total dissolved solids, nitrate nitrogen, phosphate phosphorus, phytoplankton, carbonaceous BOD, ammonia nitrogen, and dissolved oxygen.

HEC utility programs. Several utility computer programs are available from the USACE Hydrologic Engineering Center, including GEDA, HEATX, and WEATHER, which are intended for use with the WQRRS and HEC-5Q packages. The Geometric Elements from Cross Section Coordinates (GEDA) program serves as a preprocessor for WQRRS (SHP) and HEC-5Q, which prepares tables of hydraulic elements from HEC-2 (see Chapter 8) cross-sections. The Heat Exchange Program (HEATX) is used to analyze day-to-day variations in meteorologic variables and compute equilibrium temperatures and coefficients of surface heat exchange between a water surface and the atmosphere. HEATX outputs coefficients required for the WQRRS and HEC-5Q models. Program WEATHER was developed to assist users of WQRRS and HEC-5Q with the preparation of required weather input data. WEATHER reads a NOAA National Climatic Center weather data file and outputs a file in the proper input format for either WQRRS or HEC-5Q.

One-Dimensional Models for Reservoirs and Lakes

A number of models provide capabilities for simulating the vertical distribution of thermal energy and chemical and biological materials in a reservoir through time. These models include LAKECO (Chen et al. 1975), WRMMS (Tennessee Valley Authority 1976), DYRESM (Imberger et al. 1978), RESTEMP (Brown and Shiao 1981), USGS model (House 1981), MS CLEAN (Park et al. 1981), RESQUALII (Stefan et al. 1982), and MINLAKE (Riley and Stefan 1987) as well as the SELECT and CE-QUAL-R1 models described next.

SELECT. The USACE Waterways Experiment Station model SELECT (Davis et al. 1987) predicts the vertical extent and distribution of withdrawal from a reservoir of known density and quality distribution for a given discharge from a specified location. Using this prediction for the withdrawal zone, SELECT computes the quality of the release for parameters (such as temperature, dissolved oxygen, and iron) treated as conservative substances. SELECT can be used as a stand-alone program but has also been incorporated in subroutine form into other models including CE-QUAL-R1.

CE-QUAL-R1. CE-QUAL-R1 (USACE Waterways Experiment Station 1986a) is included in the Model Inventory Appendix. CE-QUAL-R1 determines values for water quality parameters as a function of vertical location and time. A reservoir is conceptualized as a vertical sequence of horizontal layers with thermal energy and materials uniformly distributed in each layer. The primary physical processes modeled include surface heat transfer, shortwave and longwave radiation and penetration, convective mixing, wind- and flow-induced mixing, entrainment of ambient water by pumped-storage inflows, inflow density current placement, selective withdrawal, and density stratification as impacted by temperature and dissolved and suspended solids. Chemical and biological processes simulated by CE-QUAL-R1 include the effects on dissolved oxygen of atmospheric exchange, photosynthesis, respiration, organic matter decomposition, nitrification, and chemical oxidation of reduced substances; uptake, excretion, and regeneration of phosphorus and nitrogen and nitrification-denitrification under aerobic and anaerobic conditions; carbon cycling and alkalinity-pH-CO_2 interactions; trophic relationships for phytoplankton and macrophytes; transfers through higher trophic levels; accumulation and decomposition of detritus and organic sediment; coliform bacteria mortality; and accumulation and reoxidation of manganese, iron, and sulfide when anaerobic conditions prevail. Reservoir outflows may be based optionally on (1) a user-specified schedule of port releases or (2) model-selected port releases to meet user-specified release amounts and temperature.

Multidimensional Models for Reservoirs and Lakes

Two-dimensional models are significantly more complex than one-dimensional models and have been used less frequently. However, in long, deep reservoirs, both vertical and longitudinal water quality gradients may be important. Three-dimensional water quality models have been used much less frequently than two-dimensional models. No generalized operational three-dimensional models were identified for inclusion in the Model Inventory Appendix. The models reported in the literature are complex and are not operational in the sense of the one- and two-dimensional models included in the Model Inventory Appendix. Examples of three-dimensional models include those reported by Simons (1973), Leendertse and Liu (1975), Funkquist and Gidhapen (1984), and Thomann et al. (1979). Several two-dimensional models are cited next.

The Box Exchange Transport Temperature and Ecology of Reservoirs (BETTER) model has been applied to Tennessee Valley Authority Reservoirs (Brown

1985). BETTER incorporates a modeling approach in which the reservoir is segmented into an array of volume elements or boxes. The flow patterns of the reservoir are modeled as longitudinal and vertical flow transfers between the array of volume elements.

The Computation of Reservoir Stratification (COORS) model (Tennessee Valley Authority 1986; Waldrop et al. 1980) and the Laterally Averaged Reservoir Model (LARM) (Buchak and Edinger 1984; Edinger and Buchak 1983) solve advection/diffusion equations in a vertical-longitudinal plane through a reservoir. COORS and LARM both provide capabilities for predicting the temperature structure of deep reservoirs throughout the annual stratification cycle. The models also develop the temporal and spatial hydrodynamics of reservoirs to provide advective components for water quality models. CE-QUAL-W2 was developed by expanding LARM to include 20 water quality constituents.

CE-QUAL-W2. The USACE Waterways Experiment Station (1986b) model CE-QUAL-W2 is included in the Model Inventory Appendix. CE-QUAL-W2 is a numerical, two-dimensional, laterally averaged model of hydrodynamics and water quality. The model consists of directly coupled hydrodynamic and water quality transport models. Hydrodynamic computations are influenced by variable water density caused by temperature, salinity, and dissolved and suspended solids. The model was developed primarily for reservoirs but can also be applied to rivers and narrow stratified estuaries.

The physical, chemical, and biological processes of CE-QUAL-W2 are similar to those of the previously discussed CE-QUAL-R1, with the following exceptions. CE-QUAL-W2 does not include transfer to higher trophic levels of zooplankton and fish. It does not account for substances accumulated in the sediments other than organic matter. It contains only one algal group rather than three. It does not include macrophytes. It does not allow the release and oxidation of sulfur and manganese when anaerobic conditions prevail, although it does allow specification, as a boundary condition, of flux from the sediments of iron, ammonia nitrogen, and phosphate phosphorus during anaerobic conditions.

COMPARISON OF MODELS INCLUDED IN THE MODEL INVENTORY APPENDIX

The following seven models are included in the Model Inventory Appendix:

QUAL2E: Enhanced Stream Water Quality Model—EPA Center for Exposure Assessment Modeling

WASP: Water Quality Analysis Program—EPA Center for Exposure Assessment Modeling

CE-QUAL-RIV1:   Dynamic One-Dimensional Model for Streams—USACE Waterways Experiment Station

CE-QUAL-R1:   Numerical One-Dimensional Model of Reservoir Water Quality—USACE Waterways Experiment Station

CE-QUAL-W2:   Two-Dimensional, Laterally Averaged Model of Hydrodynamics and Water Quality—USACE Waterways Experiment Station

WQRRS: Water Quality for River-Reservoir Systems—USACE Hydrologic Engineering Center

HEC-5Q: Simulation of Flood Control and Conservation Systems (Water Quality Version)—USACE Hydrologic Engineering Center.

The selection and use of models for investigating river and reservoir water quality problems is governed by the goals and objectives of the study. The choice of model depends on the questions to be answered by the simulation exercises. The appropriateness of modeling assumptions is also an important consideration in selecting a model for a particular application. The ability to collect the proper data to define boundary and initial conditions and to calibrate and verify the model is particularly important in water quality modeling. Limited data collection resources can constrain the choice of model.

QUAL2E and CE-QUAL-R1 are the simplest of the seven models. QUAL2E is a steady-state, one-dimensional (longitudinal) model for analyzing water quality along the length of rivers and reservoirs subject to pollutant loadings. CE-QUAL-R1 simulates the one-dimensional vertical distribution of water quality parameters in a reservoir as a function of time. CE-QUAL-RIV1 is more complex than QUAL2E in that capabilities are provided for modeling dynamic (time varying) conditions in regulated rivers. CE-QUAL-W2 is more complex than CE-QUAL-R1 in that two-dimensional modeling capabilities are provided. WASP, WQRRS, and HEC-5 are relatively sophisticated packages for modeling complex water quality problems in reservoir and river systems. WASP, WQRRS, and HEC-5 are each designed to provide a flexible general framework for addressing a broad range of modeling applications.

REFERENCES

AMBROSE, R. B., JR., et al. "WASP4, a Hydrodynamic and Water Quality Model—Model Theory, User's Manual, and Programmer's Guide," EPA-600-3-86-058, U.S. Environmental Protection Agency, Athens, Georgia, 1988.

AMBROSE, R. B., JR., and S. VANDERGRIFT, "SARAH, a Surface Water Assessment Model for Back Calculating Reductions in Abiotic Hazardous Wastes," Report EPA/600/3-86/058, Assessment Branch, Environmental Research Lab., U.S. Environmental Protection Agency, Athens, Georgia, 1986.

AMBROSE, R. B., T. A. WOOL, J. L. MARTIN, J. P. CONNOLLY, and R. W. SCHANZ, "WASP5.x, a Hydrodynamic and Water Quality Model—Model Theory, User's Manual, and Programmer's Guide," Environment Research Laboratory, U.S. Environmental Protection Agency, Draft, 1993.

BAUER, D. P., M. E. JENNINGS, and J. E. MILLER, "One-Dimensional Steady-State Stream Water Quality Model," U.S. Geological Survey Water Resources Investigation 79-45, Bay St. Louis, Mississippi, 1979.

BAUER, D. P., and N. YOTSUKURA, "Two-Dimensional Excess Temperature Model for a Thermally Loaded Stream," Report WRD-74-044, U.S. Geological Survey, Water Resources Division, Gulf Coast Hydroscience Center, NSTL Station, Mississippi, 1974.

BEDFORD, K. W., R. M. SYKES, and C. LIBICKI, "Dynamic Advective Water Quality Model for Rivers," *Journal of Environmental Engineering Division,* American Society of Civil Engineers, Vol 109, No. 3, 1983.

BROWN, R. T. "BETTER, a Two-Dimensional Reservoir Water Quality Model: Model User's Guide," Center for the Management, Utilization, and Protection of Water Resources, Tennessee Technological University, Cookeville, Tennessee, 1985.

BROWN, R. T., and M. C. SHIAO, "Reservoir Temperature Modeling: Case Study of Normandy Reservoir," Report WR28-1-86-101, Tennessee Valley Authority, Water Systems Development Branch, 1981.

BROWN, R. T., and T. O. BARNWELL, "The Enhanced Stream Water Quality Models QUAL2E and QUAL2E-UNCAS: Documentation and User Manual," EPA/600/3-87/007, U.S. Environmental Protection Agency, Environmental Research Laboratory, Athens, Georgia, May 1987.

BUCHAK, E. M., and J. E. EDINGER, "Generalized, Longitudinal-Vertical Hydrodynamics and Transport: Development, Programming and Applications," Document No. 84-18-R, U.S. Army Corps of Engineers, Vicksburg, Mississippi, 1984.

CHAPRA, S. C., and K. H. RECKHOW, *Engineering Approaches for Lake Management,* 2 volumes, Butterworth Publishers, 1983.

CHEN, C. W., M. LORENZEN, and D. J. SMITH, "A Comprehensive Water Quality-Ecologic Model for Lake Ontario," Great Lakes Environmental Research Laboratory, National Oceanic and Atmospheric Administration, Washington, D.C., 1975.

DAVIS, J. E., J. P. HOLLAND, M. L. SCHNEIDER, and S. C. WILHELMS, S. C. "SELECT: A Numerical One-Dimensional Model for Selective Withdrawal," Instruction Report E-87-2, U.S. Army Engineer Waterways Experiment Station, Vicksburg, Mississippi, 1987.

DITORO, D. M., et al. "Documentation for Water Quality Analysis Simulation Program (WASP) and Model Verification Program (MVP)," EPA-600/3-81-044, U.S. Environmental Protection Agency, Duluth, Minnesota, 1983.

EDINGER, J. E., and E. M. BUCHAK, "Developments in LARM2: A Longitudinal-Vertical,

Time-Varying Hydrodynamic Reservoir Model," Technical Report E-83-1, U.S. Army Engineer Waterways Experiment Station, Vicksburg, Mississippi, 1983.

FUNKQUIST, L., and L. GIDHAPEN, "A Model for Pollution Studies in the Baltic Sea," Report RHO 39, Swedish Meteorological and Hydrological Institute, Norkoping, Sweden, 1984.

GOWDA, T. P., "Water Quality Prediction in Mixing Zones of Rivers," *Journal of Environmental Engineering,* Vol. 110, No. 4, 1984.

HOUSE, L. B. "One-Dimensional Reservoir-Lake Temperature and Dissolved Oxygen Model," U.S. Geological Survey, Water-Resources Investigations, No. 82-5, Gulf Coast Hydroscience Center, NSTL Station, Mississippi, 1981.

IMBERGER, J., J. PATTERSON, B. HERBERT, and L. LOH, "Dynamics of Reservoirs of Medium Size," *Journal of Hydraulic Engineering,* American Society of Civil Engineers, Vol. 104, No. 4, 1978.

KRISHNAPPAN, B. G., and Y. L. LAU, "User's Manual: Prediction of Transverse Mixing in Natural Streams Model—RIVMIX MK2," Environmental Hydraulics Section, Hydraulics Division, National Water Research Institute Canada Centre for Inland Waters, Ontario, Canada, 1985.

LEENDERTSE, J. J., and S. K LIU, "A Three-Dimensional Model for Estuaries and Coastal Seas: Aspects of Computation," R-1764-OWRT, U.S. Department of the Interior, Washington, D.C., 1975.

MARTIN, J. L., "Simplified, Steady-State Temperature and Dissolved Oxygen Model: User's Guide," Instruction Report E-86-4, U.S. Army Engineer Waterways Experiment Station, Vicksburg, Mississippi, 1986.

MCCUTCHEON, S. C., *Water Quality Modeling, Volume I Transport and Surface Exchange in Rivers,* CRC Press, 1989.

ORLOB, G. T., "Water-Quality Modeling for Decision Making," *Journal of Water Resources Planning and Management,* ASCE, Vol. 118, No. 3, May 1992.

ORLOB, G. T., (editor), " Water Quality Modeling, Streams, Lakes, and Reservoirs," IIASA State of the Art Series, Wiley Interscience, London, U.K., 1984.

PARK, R. A., C. D. COLLINS, D. K. LEUNG, C. W. BOYDEN, J. ALBANESE, P. DECAPPARIIS, and H. FORSTNER, "The Aquatic Ecosystem Model MS CLEANER." Proceedings of the First International Conference on State of the Art of Ecological Modeling, S. E. Jorgensen, editor, Elsevier Publishing Co., Copenhagen, Denmark, 1981.

RILEY, M. J., and H. G. STEFAN, "Dynamic Lake Water Quality Simulation Model MIN-LAKE," Report 263, St. Anthony Falls Hydraulic Laboratory, University of Minnesota, Minneapolis, Minnesota, 1987.

ROESNER, L. A., J. R. MONSER, and D. E. EVENSON, "Computer Program Documentation for the Stream Quality Model, QUAL-II," U.S. Environmental Protection Agency, Washington, D.C., 1973.

SIMONS, T. J., "Development of Three-Dimensional Numerical Models of the Great Lakes," Scientific Series No. 12, Canada Centre for Inland Waters, Burlington, California, 1973.

STEFAN, H. G., J. J. CARDONI, and A. Y. FU, "RESQUAL II: A Dynamic Water Quality Simulation Program for a Stratified Shallow Lake or Reservoir: Application to Lake Chicot,

Arkansas," Report No. 209, St. Anthony Falls Hydraulic Laboratory, University of Minnesota, Minneapolis, Minnesota, 1982.

STEFAN, H. G., R. B. AMBROSE, and M. S. DORTCH, "Formulation of Water Quality Models for Streams, Lakes, and Reservoirs: Modeler's Perspective," Miscellaneous Paper E-89-1, U.S. Army Waterways Experiment Station, Vicksburg, Mississippi, July 1989.

TCHOBANOGLOUS, G., and E. D. SCHROEDER, *Water Quality,* Addison-Wesley, 1985.

TENNESSEE VALLEY AUTHORITY, "Water Temperature Prediction Model for Deep Reservoirs (WRMMS)," Report A-2, Norris, Tennessee, 1976.

TENNESSEE VALLEY AUTHORITY, "Hydrodynamic and Water Quality Models of the TVA Engineering Laboratory," Issue No. 3, Norris, Tennessee, September 1986.

TEXAS WATER DEVELOPMENT BOARD, "Simulation of Water Quality in Streams and Canals: Theory and Description of QUAL-I Mathematical Modeling System," Report 128, Austin, Texas, 1971.

THOMANN, R. V., "Mathematical Model for Dissolved Oxygen," *Journal of Sanitary Engineering,* ASCE, Vol. 89, No. 5, 1963.

THOMANN, R. V., and J. A. MUELLER, *Principles of Surface Water Quality Modeling and Control,* Harper & Row, New York, 1987.

THOMANN, R. V., R. P. WINFIELD, and J. J. SEGNA, "Verification Analysis of Lake Ontario and Rochester Embayment Three-Dimensional Eutrophication Models," Report No. EPA-600/3-79-094, U.S. Environmental Protection Agency, Duluth, Minnesota, August 1979.

U.S. ARMY CORPS OF ENGINEERS, HYDROLOGIC ENGINEERING CENTER, "Water Quality for River-Reservoir Systems (WQRRS), User's Manual," CPD-8, Davis, California, October 1978, revised February 1985.

U.S. ARMY CORPS OF ENGINEERS, HYDROLOGIC ENGINEERING CENTER, "HEC-5 Simulation of Flood Control and Conservation Systems, Appendix on Water Quality Analysis," Davis, California, Draft, September 1986.

U.S. ARMY CORPS OF ENGINEERS, WATERWAYS EXPERIMENT STATION, "CE-QUAL-R1: A Numerical One-Dimensional Model of Reservoir Water Quality; User's Manual," Instruction Report E-82-1, Environmental Laboratory, Vicksburg, Mississippi, July 1986a.

U.S. ARMY CORPS OF ENGINEERS, WATERWAYS EXPERIMENT STATION, "CE-QUAL-W2: A Numerical Two-Dimensional, Laterally Averaged Model of Hydrodynamics and Water Quality; User's Manual," Instruction Report E-86-5, Environmental and Hydraulics Laboratories, Vicksburg, Mississippi, August 1986b.

U.S. ARMY CORPS OF ENGINEERS, WATERWAYS EXPERIMENT STATION, "CE-QUAL-RIV1: A Dynamic One-Dimensional (Longitudinal) Water Quality Model for Streams; User's Manual," Instruction Report E-90-1, Environmental Laboratory, Vicksburg, Mississippi, November 1990.

WALDROP, W. R., C. D. UNGATE, and W. L. HARPER, "Computer Simulation of Hydrodynamics and Temperatures of Tellico Reservoir," Report No. WR28-1-65-100, TVA Water Systems Development Branch, Norris, Tennessee, 1980.

10

RESERVOIR/RIVER SYSTEM
OPERATION MODELS

INTRODUCTION

This chapter covers simulation and optimization models for analyzing water quantity aspects of reservoir system operations. Water quality considerations were addressed in Chapter 9. Chapter 10 is the only chapter in which optimization (mathematical programming) models are emphasized as well as simulation models. Optimization techniques are also applied in the modeling areas addressed by the other chapters, but not to the extent as in the analysis of reservoir/river system operations.

Model Applications

Reservoir/river system analysis models are used for various purposes in a variety of settings. Models are used in planning studies to aid in the formulation and evaluation of alternative plans for responding to water-related problems and needs. Feasibility studies may involve proposed construction projects as well as reallocations of storage capacity or other operational modifications at completed projects. Another modeling application involves studies made specifically to reevaluate operating policies for existing reservoir systems. Periodic reevaluations may be made routinely to assure system responsiveness to current conditions and objectives. However, more typically, reevaluation studies are made in response to a particular perceived problem or need. Studies may be motivated by the existence of severe drought conditions.

Development of drought contingency plans, in preparation for future droughts, is an important activity which is receiving increasing attention. Execution of models during actual reservoir operations in support of real-time release decisions represents another major area of application.

Reservoir System Operating Plans

Reservoir system management practices and associated modeling and analysis methods involve

- Allocating storage capacity and streamflow between multiple uses and users
- Minimizing the risks and consequences of water shortages and flooding
- Optimizing the beneficial use of water, energy, and land resources
- Managing environmental resources

A reservoir regulation plan, operating procedure, or release policy is a set of rules for determining the quantities of water to be stored and to be released or withdrawn from a reservoir or system of several reservoirs under various conditions. Typically, a regulation plan includes a set of quantitative criteria within which significant flexibility exists for qualitative judgment. In real-world operations, the operating rules provide guidance to the water managers who make the actual release decisions. In modeling exercises, the reservoir system analysis model contains some mechanism for making period-by-period release decisions within the framework of user-specified operating rules and/or criteria functions.

Reservoir/River System Analysis Literature

A tremendous amount of work on developing and applying reservoir/river system operation models has been documented in the published literature during the past several decades. Much additional work has been accomplished without being reported in the published literature. Yeh (1985) provides a comprehensive, state-of-the-art review of reservoir operation models with a strong emphasis on optimization methods. Wurbs et al. (1985) provide an annotated bibliography which cites several hundred references on reservoir systems analysis models. The USACE Hydrologic Engineering Center (1991c) presents a comprehensive review of reservoir system modeling and analysis approaches. Wurbs (1993) also reviews reservoir system simulation and optimization models.

Model Categorization

Reservoir/river system analysis models have traditionally been categorized as (1) simulation, (2) optimization, and (3) combinations of simulation and optimization. In a broad sense, optimization includes human judgment, use of either simulation and/or optimization models, and use of other decision support tools. However, following common usage in the literature, the term *optimization* is used here, synonymously with *mathematical programming,* to refer to a mathematical formulation in which a formal algorithm is used to compute a set of decision variable values which minimize or maximize an objective function subject to constraints. Whereas optimization models automatically search for an "optimum" set of decision variable values, simulation models are limited to predicting system performance for a user-specified set of variable values. A simulation model is a representation of a system used to predict the behavior of the system under a given set of conditions. Alternative runs of a simulation model are made to analyze the performance of the system under varying conditions, such as for alternative operating policies.

Although optimization and simulation are two alternative modeling approaches with different characteristics, the distinction is somewhat obscured by the fact that most models, to various degrees, contain elements of both approaches. All optimization models also simulate the system. Optimization algorithms are embedded within many major reservoir system simulation models to perform certain computations. An optimization approach may involve numerous iterative executions of a simulation model, with the iterations being automated to various degrees. Simulation and optimization models can also be used in combination. For example, a study could involve preliminary screening of numerous alternatives using an optimization model, followed by a more detailed evaluation of selected plans using a simulation model.

Development and application of decision support tools within the federal water resources development agencies have focused on simulation models. The academic community and research literature have emphasized optimization techniques, particularly linear and dynamic programming but other nonlinear programming methods as well. Reservoir system analysis models based on network flow programming provide a particularly notable example of combining advantageous features of simulation with a formal optimization algorithm. In recent years, a major emphasis in all modeling sectors, regardless of the simulation or optimization computational algorithms being used, has been to use desktop computers as well as larger computers and to take advantage of advances in computer technology in the areas of data management, output analysis, graphics, and user-friendly human-machine interfacing.

Reservoir/river system analysis models have traditionally been developed using FORTRAN or similar programming languages. In recent years, models have also been developed using the object-oriented programming approach, typically with object-oriented versions of the C language. As noted in Chapter 1, instructions and information are coded and stored as objects or modules in this model development environment. Objects can be reused in different programs and subprograms, and models are easier to modify. Graphical user interfaces and data management software are typically combined with object-oriented simulation models. These modeling packages based on object-oriented programming are typically implemented on workstations.

Reservoir/river system operation models are grouped as follows for the following review:

- Simulation models developed for specific reservoir systems
- Optimization models
- Generalized reservoir system simulation models
- Generalized reservoir system analysis models based on network flow programming
- Generalized reservoir system analysis models based on object-oriented programming.

REVIEW OF AVAILABLE MODELS

Simulation Models Developed for Specific Reservoir/River Systems

A reservoir system simulation model reproduces the hydrologic and, in some cases, economic performance of a reservoir/stream system for given hydrologic inputs (streamflows and reservoir evaporation rates) and operating rules. The models are based on mass-balance accounting procedures for tracking the movement of water through a reservoir/stream system. Various strategies can be adopted for applying simulation models. Series of runs are typically made to compare system performance for alternative reservoir configurations, storage allocations, operating rules, demand levels, and/or hydrologic inflow sequences. System performance may be evaluated by simply observing the computed time sequences of storage levels, releases, hydroelectric power generated, water supply diversions and diversion shortages, and instream flows. Various types of storage or discharge frequency analyses may be performed. Simulation models may also provide the capability to analyze reservoir system operations using hydrologic and economic performance measures such as firm yield, yield versus reliability relationships, hydroelectric revenues, flood damages, and economic benefits and costs associated with various purposes.

Simulation modeling of major river basins began in the United States in 1953 with a study by the USACE of the operation of six reservoirs on the Missouri River. The objective was to maximize hydroelectric power generation subject to constraints imposed by specified requirements for navigation, flood control, and irrigation. Simulation models for numerous other river/reservoir systems have been developed since then. (A recent study of the Missouri River system is discussed later in this chapter.) Early simulation models as well as many of the more recently developed models are for specific reservoir systems. Descriptions of many site-specific simulation models can be found in the published literature. Numerous other models successfully used in offices throughout the United States have never been reported in the literature. The U.S. Bureau of Reclamation (1991) presents descriptions of a number of simulation models used at specific Bureau of Reclamation projects. The Missouri, Potomac, and Colorado River models discussed next are particularly notable examples of site-specific reservoir/river system simulation models.

In 1989, the Corps of Engineers initiated a several-year study to review the *Missouri River Master Water Control Manual,* which guides operation of the six mainstem Missouri River reservoirs located in Montana, North Dakota, South Dakota, and Nebraska (Cieslik and McAllister 1994). The reservoirs are operated for flood control, hydropower, water supply, water quality, irrigation, navigation, recreation, and environmental resources. The Long Range Study (LRS) Model uses a monthly time interval for simulating operation of the system during 96-year sequences of historical flows at six reservoir nodes and six other gauge locations (Patenode and Wilson 1994). The original model developed in the 1960s was updated during the 1990s and modified for execution on IBM-compatible desktop computers. The model has been used to simulate and evaluate numerous alternative operating policies. The results of the LRS hydrologic simulation model are used in combination with several other models, including environmental and economic analysis models, in evaluating operations of this reservoir system.

The Potomac River Interactive Simulation Model (PRISM) was originally developed by a research team at Johns Hopkins University (Palmer et al. 1982). A number of water management agencies in the Potomac River Basin participated in drought simulation exercises using PRISM during development and implementation of a regional water supply plan for the Washington Metropolitan Area. The Corps of Engineers modified PRISM for use in certain drought simulation studies (USACE Baltimore District 1983). PRISM simulates the operation of the four reservoirs and allocation of water within the Washington Metropolitan Area. Versions of the model alternatively use a weekly and daily time interval. The model determines the amount of water available to each of the several jurisdictions for given streamflows, demands, and water allocation and reservoir operation rules. When operating in the batch mode, PRISM performs the functions of a regional water supply manager in strict accordance with rules specified by the model user. The interactive model allows participants to engage in a dialogue with the model as it is being executed, thereby changing model parameters and overriding prespecified decision rules. The interactive model represents an attempt to include, in a formal analytical modeling exercise, the process by which water supply management decisions are made.

The Colorado River Simulation System (CRSS), originally developed by the Bureau of Reclamation during the 1970s and subsequently revised and updated, simulates operations of the major reservoirs in the Colorado River Basin for water supply, low-flow augmentation, hydroelectric power, and flood control (Schuster 1987). The CRSS is a set of computer programs and databases used in long-range planning. The monthly time-interval historical hydrologic period-of-record model reflects operation of the system in accordance with a series of river basin compacts, laws, and agreements collectively called the law of the river. Salt concentrations are also considered.

Optimization Models

During World War II, the Allies organized interdisciplinary teams to solve complex scheduling and allocation problems involved in military operations. Mathematical optimization models were found to be very useful in this work. After the war, the evolving discipline of operations research continued to rely on optimization models for solving a broad range of problems in private industry. The same mathematical programming techniques also became important tools in the various systems engineering disciplines, including water resources systems engineering. Reservoir operations have been viewed as an area of water resources planning and management that has particularly high potential for beneficial application of optimization models.

The literature related to optimization models in general and application to reservoir operations in particular is extensive. Mathematical programming is covered by numerous operations research and mathematics books as well as by water resources system analysis textbooks by Loucks et al. (1981) and Mays and Tung (1992). Yeh (1985) presents a comprehensive, in-depth, state-of-the-art review of reservoir operation models, with a strong emphasis on optimization techniques.

Numerous applications of optimization techniques to reservoir operation problems have been reported in the published literature during the past several decades. With the notable exception of the generalized network flow models (discussed later in this chapter), the models have usually been developed for a specific reservoir system. University research projects involving case studies account for most of the applications of optimization techniques to reservoir operations to date. Most of the applications involve various variations and extensions of linear programming and dynamic programming. Network flow programming, discussed later, is a special form of linear programming. Various other nonlinear programming methods have been used to a lesser extent.

Optimization models are formulated in terms of determining values for a set of decision variables which will maximize or minimize an objective function subject to constraints. The objective function and constraints are represented by mathematical expressions as a function of the decision variables. For a reservoir operation problem, the decision variables are typically release rates and end-of-period storage volumes. Constraints typically include storage capacities and other physical characteristics of the reservoir/stream system, diversion or instream flow requirements for various purposes, and mass balances. In the multiple-stage optimization approach of dynamic programming, stages often represent time periods tied together by the state variable of storage.

Objectives and objective functions. The objective function is the heart of an optimization model. The objective function, which is to be minimized or maximized subject to constraints, must be expressed in the proper mathematical format as a function of the decision variables. For example, in linear programming, the solution algorithm finds values for the n decision variables (X_j) which minimize the linear objective function (Z):

$$Z = c_1 X_1 + c_2 X_2 + c_3 X_3 + \ldots + c_n X_n$$

where the c_j are constants.

The objective function may be a penalty or utility function (such as the network flow programming approach of defining operating rules based on relative priorities, as discussed later) or may be a mathematical expression of a planning or operational objective, such as the examples listed next. The following objectives have been reflected in the objective functions of various optimization models reported in the literature:

- Economic Benefits and Costs
  - Maximize water supply and/or hydroelectric power revenues.
  - Minimize the cost of meeting electric power commitments from a combined hydro/thermal system.
  - Minimize economic losses due to water shortages.
  - Minimize the electrical cost of pumping water in a distribution system.
  - Minimize the damages associated with a specified flood event.
  - Maximize the net benefits of multiple-purpose operations.
  - Minimize costs associated with multiple-purpose operations.

- Water Availability and Reliability
  - Maximize firm yield, yields for specified reliabilities, or reliabilities for specified demands.
  - Minimize shortage frequencies and/or volumes.
  - Minimize shortage indexes, such as the sum of the squared deviations between target and actual diversions.
  - Minimize the weighted sum of shortage indexes.
  - Maximize the minimum streamflow.
  - Maximize reservoir storage at the end of the optimization horizon.
  - Minimize spills.
  - Minimize evaporation losses.
  - Minimize average monthly storage fluctuations.
  - Maximize the length of the navigation season.
  - Minimize the total volume of water released for minimum navigation needs.
- Hydroelectric Power Generation
  - Maximize firm energy.
  - Maximize average annual energy.
  - Minimize energy shortages or energy shortage indexes.
  - Maximize the potential energy of water stored in a system.

Although several different objectives will typically be of concern in a particular reservoir system analysis study, an optimization model can incorporate only one objective function. Multiple objectives can be combined in a single function if they can be expressed in commensurate units, such as dollars. However, objectives are often not quantified in commensurate units. Two alternative approaches are typically adopted to analyze trade-offs between objectives.

One approach is to execute the optimization model with one selected objective reflected in the objective function and the other objectives treated as constraints at fixed user-specified levels. For example, the model might maximize average annual energy subject to the constraint that a user-specified water supply firm yield and firm energy be maintained. Alternative runs of the model could be made to show how the average annual energy is affected by changes in the user-specified water supply firm yield and firm energy.

An alternative approach to analyzing trade-offs between noncommensurate objectives involves treating each objective as a weighted component of the objective function. The objective function is the sum of each component multiplied by a weighting factor reflecting the relative importance of that objective. The weighting factors can be arbitrary, with no physical significance other than to reflect relative weights assigned to the alternative objectives included in the objective function. The model can be executed iteratively, with different sets of weighting factor values to analyze the trade-offs between the objectives with alternative operating plans.

Linear programming. Of the many optimization techniques, linear programming (LP) is the simplest and most widely used in many fields. Its popularity in water resources systems analysis as well as in other operations research, management science, and systems engineering disciplines is due to the following factors: (1) LP is applicable to a wide variety of types of problems; (2) efficient solution algorithms are available; and (3) generalized computer software packages are readily available for applying the solution algorithms.

LP consists of minimizing or maximizing a linear objective function subject to a set of linear constraints. Nonlinearities are associated with various features of reservoir operation models, such as evaporation and hydroelectric power computations and benefit/cost functions. In some cases, nonlinear features of a problem may be approximated by linearization techniques for incorporation into an LP model. Successive iterative solutions of an LP problem are sometimes used to handle nonlinearities.

Considerable ingenuity and significant approximations may be required to formulate a real-world problem in the required mathematical format. However, if the problem can be properly formulated, standard LP algorithms and computer codes are available to perform the computations. Special computationally efficient algorithms are available for certain forms of LP problems, such as the network flow programming models discussed later.

Dynamic programming. Dynamic programming (DP) is not a precise algorithm like LP, but a general approach to solving optimization problems. DP involves decomposing a complex problem into a series of simpler subproblems which are solved sequentially while transmitting essential information from one stage of the computations to the next using state concepts. DP models have the following characteristics:

1. The problem is divided into stages, with a decision required at each stage. The stages may represent different points in time (as in determining reservoir releases for each time interval), different points in space (e.g., releases from different reservoirs), or different activities (such as releases for different water users).

2. Each stage of the problem must have a definite number of states associated with it. The states describe the possible conditions in which the system might be at that stage of the computations. The amount of water in storage is an example of a typical state variable.

3. The effect of a decision at each stage of the problem is to transform the current state of the system into a state associated with the next stage. If the decision variable is how much water to release from the reservoir at the current time, this decision will transform the amount of water stored in the reservoir (state variable) from the current amount to a new amount for the next stage (time period).

4. A return function that indicates the utility or cost of the transformation is associated with each potential state transformation. The return function allows the objective function to be represented by stages.

5. The optimality of the decision required at the current stage is judged in terms of its impact on the return function for the current stage and all subsequent stages.

Loucks et al. (1981) and Mays and Tung (1992) describe the fundamentals of DP. Yeh (1985) reviews the various types and variations of DP formulations of reservoir operation models. Labadie (1990) describes a generalized microcomputer DP package, called CSUDP, which has been used for a broad range of water resources planning and management applications.

DP is a general methodology for applying state concepts to the formulation and solution of problems which can be viewed as optimizing a multiple-stage decision process. In some cases, the same problem can be solved by alternative DP formulations or by either DP or LP. In general, LP has the advantage over DP of being a more precisely defined, easier to understand algorithm. The degree of generalization and availability of generalized computer codes is much more limited for DP than for LP. However, the strict linear form of the LP formulation can be a significant hindrance. Nonlinear properties of a problem can be reflected readily in a DP formulation. However, various assumptions, including a separable objective function, limit the range of application and require ingenuity and understanding by the modeler in applying DP. The so-called curse of dimensionality is a major consideration in DP, which means that increasing the number of state variables greatly increases the computational burden. For example, because reservoir storage is typically a state variable, the number of reservoirs which can be included in a DP model may be limited.

Other nonlinear programming algorithms. Nonlinear programming (NLP) algorithms, such as quadratic programming, geometric programming, and separable programming, are covered in standard operations research and mathematics textbooks. These NLP methods provide a more general mathematical formulation than LP, but the mathematics involved are much more complicated. The NLP techniques have been applied relatively little, compared to LP and DP, to problems of optimizing reservoir operations. The significant advancements in computer technology in recent years have removed computational constraints, which could result in greater use of NLP in the future.

Next to LP and DP, search algorithms represent the optimization approach which has been used most extensively in optimizing reservoir operations. Search methods iteratively change the values of the decision variables in such a way as to move closer to the optimum value of the objective function. A broad range of nonlinear optimization techniques is classified as search methods. Search techniques are typically combined with simulation models and have also been combined with LP and DP.

Examples of optimization models.   Yeh (1985) and Wurbs et al. (1985) provide extensive lists of references on the use of linear programming, dynamic programming, and other nonlinear programming methods in analyses of reservoir operations. Several models are discussed next as a representative sampling of the variety of ways in which optimization techniques have been applied. The examples cited include both models developed for specific reservoir systems and generalized models which can be applied to any system.

Shane and Gilbert (1982) and Gilbert and Shane (1982) describe a model, called HYDROSIM, used to simulate the 42-reservoir Tennessee Valley Authority system based on an established set of operating priorities. A series of operating constraints is formulated to represent the various objectives. The model sequentially minimizes the violation of the constraints in their order of priority. The HYDROSIM model uses LP to compute reservoir storages, releases, and hydroelectric power generation for each week of a 52-week period beginning at the present based on alternative sequences of historical streamflows. A search procedure is used to handle a nonlinear hydropower cost function. Giles and Wunderlich (1981) describe a similar model developed by the Tennessee Valley Authority based on DP.

A real-time optimization procedure, involving combined use of DP and LP, was developed to determine multiple-reservoir release schedules for hydroelectric power generation in the operation of the California Central Valley Project (Yeh 1981). The overall procedure optimizes, in turn, a monthly model over a period of one year, a daily model over a period of up to one month, and an hourly model for 24 hours. Output from one model (monthly ending storages or daily releases) is used as input to the next echelon model. The monthly model is a combined LP-DP formulation which computes releases and storages based on the objective of minimizing the loss of stored potential energy. Given end-of-month storage levels, the daily model uses LP to determine the daily releases for each power plant, which minimizes loss of stored potential energy in the system. The hourly model uses a combination of LP and DP to determine hourly releases for each plant, which maximizes total daily system power output.

Allen and Bridgeman (1986) applied DP to three case studies involving hydroelectric power scheduling, which included (1) optimal instantaneous scheduling of hydropower units with different generating characteristics to maximize overall plant efficiency; (2) optimal hourly scheduling of hydropower generation between two hydrologically linked power plants to maximize overall daily/weekly system efficiency; and (3) optimal monthly scheduling of hydropower generation to minimize the purchase cost of imported power supply subject to a time-of-day rate structure.

Chung and Helweg (1985) combined DP with HEC-3 in an analysis of operating policies for Lake Oroville and San Luis Reservoir, which are components of the California State Water Project. (The HEC-3 simulation model is discussed later in the section on generalized reservoir system simulation models.) HEC-3 was used to determine the amount of excess water still available for export after all system commitments were met. A DP model was then used to determine how the reservoirs should be operated to maximize the net benefits of exporting the excess water. The DP decision variables were reservoir releases in each time period, and the objective function was an expression of revenues from selling the water. Because approximations were necessary in formulation of the DP model, HEC-3 was used to check and refine the release schedules determined with the DP model.

Palmer and Holmes (1988) describe the Seattle Water Department integrated drought management expert system. An LP model is incorporated in this decision support system to determine optimal operating policies and system yield. The LP model is based on the two objectives of maximizing yield and minimizing the economic loss associated with deficits from a specified target.

Randall et al. (1990) developed an LP model to study the operation, during drought, of a metropolitan water system consisting of multiple reservoirs, ground water, treatment plants, and distribution facilities. Four objectives were incorporated in the modeling study: (1) Maximize net revenues, which were the difference between revenues for selling water and electrical pumping costs; (2) maximize reliability, expressed as the minimum of the ratios of consumption to demand for each water use district; (3) maximize reservoir storage at the end of the optimization horizon; and (4) maximize the minimum flow in the streams. Alternative versions of the model were formulated, with one objective being optimized as the objective function and with the other objectives being incorporated as constraints at user-specified levels. Trade-off curves were developed to show the trade-offs between the four alternative objectives.

Martin (1987) describes the MONITOR-I model developed by the Texas Water Development Board to analyze complex surface water storage and conveyance systems operated for hydroelectric power, water supply, and low-flow augmentation. Unlike the site-specific models discussed earlier, MONITOR-I is generalized for application to any system. The LP model uses an iterative successive LP algorithm to handle nonlinearities associated with hydroelectric power and other features of the model. The decision variables are daily reservoir releases, water diversions, and pipeline and canal flows. The objective function to be maximized is an expression of net economic benefits.

Simonovic (1992) describes an intelligent decision support system, called REZES, which includes a library of 11 models for performing various analyses for a single reservoir. The models utilize various simulation and optimization techniques including LP, DP, and nonlinear programming.

Generalized Reservoir/River System Simulation Models

The term *generalized reservoir system simulation model* refers to computer programs designed to be applied readily to a variety of reservoir/stream systems. The user simply develops the input data for the particular system of interest, in a specified format, and executes the model without being concerned about developing or modifying the actual computer code. A number of such models have been reported in the literature. Several models considered to be representative of the state of the art are discussed next. Three of the models (HEC-5, IRIS, and TAMUWRAP) are included in the Model Inventory Appendix.

HEC-3 and HEC-5. The HEC-3 Reservoir System Analysis for Conservation computer program simulates the operation of a reservoir for conservation purposes such as water supply, low-flow augmentation, and hydroelectric power. HEC-3 is documented by a user's manual (USACE HEC 1981) and other publications available from the USACE Hydrologic Engineering Center (HEC). Feldman (1981) provides an overview of several generalized computer programs available from the HEC, including HEC-3 and HEC-5. Various USACE offices and other entities have modified HEC-3 for different applications. HEC-3 and HEC-5, discussed next, have similar capabilities for simulating conservation operations. However, HEC-3 does not have the comprehensive flood control capabilities of HEC-5. The Hydrologic Engineering Center has not actively updated or expanded HEC-3 in recent years because essentially all of its capabilities have been duplicated in HEC-5. HEC-3 is no longer distributed by the HEC.

The HEC-5 Simulation of Flood Control and Conservation Systems computer program is probably the most versatile of the available models in the sense of being applicable to a wide range of both flood control and conservation operation problems. HEC-5 is well documented and has been used in a relatively large number of studies, including studies of storage reallocations and other operational modifications at existing reservoirs as well as feasibility studies for proposed new projects. The program is also used for real-time operation. An initial version released in 1973 has subsequently been expanded greatly. Microcomputer versions of the model have been released recently. Several utility programs are available to aid in developing input data files and analyzing output. Alternative versions of the model are available which exclude and include water quality analysis capabilities. The water quality version was discussed in Chapter 9. The HEC-5 user's manual (USACE HEC 1982, 1989) provides detailed instructions for its use. Various publications regarding the use of HEC-5 available from the Hydrologic Engineering Center include training documents covering various features of the model and reports and papers documenting specific applications of the model in actual reservoir system analysis studies. HEC-5 is also included in the inventory of short courses taught by the Hydrologic Engineering Center.

HEC-5 simulates the sequential, period-by-period operation of a multiple-purpose reservoir system for inputted sequences of unregulated streamflows and reservoir evaporation rates. The program uses a variable time interval. For example, monthly or weekly data might be used during periods of normal or low flows in combination with daily or hourly data during flood events. The user specifies the operating rules in HEC-5 by inputting reservoir storage zones, diversion and minimum instream flow targets, and allowable flood flows. The model makes release decisions to empty flood control pools and to meet user-specified diversion and instream flow targets based on computed reservoir storage levels and streamflows at downstream locations. Seasonal rule curves and buffer zones can be included in the operating rules. Multiple-reservoir release decisions are based on balancing the percent storage depletion in specified zones. Several alternative hydrologic flood routing methods are available. HEC-5 has various optional analysis capabilities, including computation of expected annual flood damages and single-reservoir firm yields for water supply and hydroelectric power.

Other USACE models. A generalized reservoir system simulation model, called SUPER, was developed by the USACE Southwestern Division (SWD) and is described by Hula (1981). SUPER performs the same types of hydrologic and economic simulation computations as HEC-5, including comprehensive flood control analyses. The SWD model uses a one-day computation interval, whereas HEC-5 uses a variable time interval. Details of handling input and output data and various computational capabilities differ somewhat between HEC-5 and SUPER. The division and district offices in the Southwestern Division have applied the model in a number of studies over a number of years. The Reservoir Modeling Center in the Tulsa District, in particular, routinely uses SUPER to simulate various major USACE reservoir systems located throughout the Southwestern Division.

The Basin Runoff and Streamflow Simulation (BRASS) model was originally developed by the USACE Savannah District to provide flood management decision support for operation of a reservoir system in the Savannah River Basin (Colon and

McMahon 1987). BRASS is an interactive hydrologic/hydraulic simulation model which includes rainfall-runoff modeling, storage routing through gated reservoirs, and dynamic streamflow routing capabilities.

The Streamflow Synthesis and Reservoir Regulation (SSARR) model was developed by the USACE North Pacific Division (NPD) primarily for streamflow and flood forecasting and reservoir operation studies. Various versions of the model date back to 1956. A program description and user's manual (USACE North Pacific Division 1987) documents the current version of the model. Various USACE and other organizations have used SSARR to model reservoir systems in the Columbia River Basin and a number of other basins. SSARR is composed of three basic components: (1) a watershed model for synthesizing runoff from rainfall and snow-melt, (2) a streamflow routing model, and (3) a reservoir regulation model. SSARR was discussed further in Chapter 7 from the perspective of its watershed modeling capabilities.

The Hydro System Seasonal Regulation (HYSSR), Hourly Load Distribution and Pondage Analysis Program (HLDPA), and Hydropower System Regulation Analysis (HYSYS) models were also developed by the USACE North Pacific Division. The models are described in the USACE Hydropower Engineer Manual (USACE OCE 1985). User's manuals are available from the NPD. HYSSR is a monthly sequential routing model designed to analyze the operation of reservoir systems for hydroelectric power and snowmelt flood control. It has been used to analyze proposed new reservoirs and operations of existing systems in the Columbia River Basin and a number of other river basins. HLDPA is an hourly time-interval planning tool designed to address such problems as optimum installed capacity, adequacy of pondage for peaking operation, and impact of hourly operation on nonpower river uses. HYSYS is a generalized model designed to support real-time operations.

MITSIM.   The MITSIM model was originally developed at the Massachusetts Institute of Technology (Strzepek et al. 1979). Early versions of the model were used in studies of the Rio Colorado in Argentina and the Vardar/Axios project in Yugoslavia and Greece. Subsequent versions have been applied in a number of studies. The model has been updated and adapted to changing computer environments at the Center for Advanced Decision Support for Water and Environmental Systems (Strzepek et al. 1989). MITSIM provides capabilities to evaluate both the hydrologic and economic performance of alternative river basin development plans involving reservoirs, hydroelectric power plants, irrigation areas, and municipal and industrial water supply diversions. A river/reservoir/use system is conceptualized as a collection of arcs and nodes. A variable computational time interval is used. The model assesses system reliability in meeting demands. Economic benefits and costs can also be evaluated. Benefits are divided into long-term benefits and short-term losses. Optional displays of net economic benefits and benefit-cost ratios for the entire river basin and/or subregions within the basin can be included in the output.

TAMUWRAP. The Water Rights Analysis Package (TAMUWRAP), developed at Texas A&M University, is designed to simulate management and use of the streamflow and reservoir storage resources of a river basin, or multiple basins, under a prior appropriation water rights permit system. The model is designed for simulation studies involving a priority-based allocation of water resources among many different water users. Water use diversions and reservoir storage facilities may be numerous and the allocation system complex. TAMUWRAP can be used to evaluate operation of a particular multiple-reservoir, multiple-user system while considering interactions with numerous other water management entities which also hold permits or rights to store and withdraw water from the river system. Multiple-reservoir release decisions are based on balancing the percent depletion in user-specified storage zones. The original model described by Wurbs and Walls (1989) has been expanded significantly (Wurbs et al. 1993). Capabilities for considering salinity have been added recently (Wurbs et al. 1994).

A monthly time interval is used, with no limit on the length of the period of analysis. Input requirements include naturalized streamflows, reservoir evaporation rates, water rights data (including permitted diversion and storage amounts, use types, return flow requirements, and priorities), hydroelectric power plant characteristics, and reservoir characteristics and operating policies. Output includes diversions, shortages, hydroelectric energy generated, streamflow depletions, unappropriated streamflows, reservoir storages and releases, and reliability statistics.

IRIS and IRAS. The Interactive River System Simulation (IRIS) model was developed with support from the Ford Foundation, United Nations Environment Program, International Institute for Applied Systems Analysis, and Cornell University (Loucks et al. 1989, 1990). Model development was motivated largely by the objective of providing a useful tool for water managers responsible for negotiating agreements among individuals and organizations in conflict over water use.

IRIS operates in a menu-driven microcomputer or workstation environment with extensive use of computer graphics for information transfer between machine and user. The configuration of the system is specified by "drawing in" nodes (reservoirs, inflow sites, junctions, and other key locations) and interconnecting links (river reaches, canals). IRIS simulates a water supply and conveyance system of essentially any normal branching configuration for inputted streamflow sequences, using a user-specified time step. Hydroelectric power and water quality features are included. System operating rules include (1) reservoir releases specified as a function of storage and season of the year, (2) allocation functions for multiple links from the same node, and (3) storage distribution targets for reservoirs operating in combination. The model has the unique feature of allowing the operating rules to be changed interactively by the user during the course of a simulation run. Another unique feature allows several alternative sets of inflow sequences to be considered in a single run of the model. Model output includes time series plots of flows, storages, energy generated, and water quality parameters at any node or link in the reservoir/river system and probability distribution displays of magnitude and duration of shortages or failure events.

The Interactive River-Aquifer Systems (IRAS) model was developed recently as an extension of IRIS (Loucks et al. 1994). IRAS is a generalized program for analyzing regional surface and ground-water management systems. IRAS is designed to be an interactive, flexible system for addressing problems involving interactions between ground and surface waters and between water quality and quantity. The model predicts the range and likelihood of various water quantity, quality, and hydropower impacts over time associated with alternative design and operating policies, for portions or entire systems, of multiple rivers and ground-water aquifers. Simulations are based on mass balances of quantity and quality constituents, taking into account flow routing, seepage, evaporation, consumption, and constituent growth, decay, and transformation, as applicable. A variable computational time step is used. IRAS provides a menu-driven, graphics-based user interface.

Generalized Reservoir/River System Analysis Models Based on Network Flow Programming

A general overview of optimization models was presented earlier in this chapter. This section focuses on a group of models which use one particular optimization technique. Network flow programming has been demonstrated to be particularly well suited to the development and practical application of generalized models. The simulation models discussed earlier are conventional simulation models in the sense that no formal mathematical programming algorithm is used. Many network flow models can also be characterized as being simulation models in the sense that they are applied in the same manner as conventional simulation models, even though the internal computations are performed differently. However, as discussed later, network flow programming also allows development of models with a more prescriptive orientation.

Network flow programming. Network flow programming is used in a broad range of operations research and systems engineering applications and is covered in a number of books, including Jensen and Barnes (1980) and Kennington and Helgason (1980). There are various recognized forms or classes of network flow problems and corresponding solution algorithms. Most reservoir systems analysis applications of network flow programming are formulated in a particular format, referred to as a minimum-cost capacitated network flow problem, which can be solved using special linear programming algorithms such as the out-of-kilter algorithm.

In a network flow model, the system is represented as a collection of nodes and arcs. For a reservoir/river system, the nodes are locations of reservoirs, diversions, stream tributary confluences, and other pertinent system features. Nodes are connected by arcs representing the way flow is conveyed. For a reservoir/river system, flow represents either a discharge rate, such as instream flows and diversions, or change in storage per unit of time.

The general form of the network flow programming problem is as follows:

minimize $\sum \sum c_{ij}q_{ij}$ for all arcs

subject to $\sum q_{ij} - \sum q_{ji} = 0$ for all nodes

and $\quad\quad l_{ij} \leq q_{ij} \leq u_{ij}$ for all arcs

where $\quad q_{ij}$ = flow rate in the arc connecting node i to node j

$\quad\quad\quad c_{ij}$ = penalty or weighting factor for q_{ij}

$\quad\quad\quad l_{ij}$ = lower bound on q_{ij}

$\quad\quad\quad u_{ij}$ = upper bound on q_{ij}

The network flow programming algorithm computes the values of the flows (q_{ij}) in each of n arcs (node i to node j), which minimizes an objective function consisting of the sum of the flows multiplied by corresponding weighting factors, subject to constraints which include maintaining a mass balance at each node and not violating user-specified upper and lower bounds on the flows. Each arc has three parameters: (1) a weighting, penalty, or unit cost factor (c_{ij}) associated with q_{ij}; (2) a lower bound (l_{ij}) on q_{ij}; and (3) an upper bound (u_{ij}) on q_{ij}. The requirement for lower and upper bounds results in the term *capacitated flow networks*. The weighting factor (c_{ij}) could be a unit cost in dollars or alternatively could be a penalty or utility term which provides a mechanism for expressing relative priorities for use in defining operating rules. A penalty weighting factor is the same as a negative utility weighting factor. Reservoir operating rules are defined by user-specified values of c_{ij}, l_{ij}, and u_{ij}. The user of the model provides lower and upper bounds on diversions, instream flows, and reservoir storage levels and assigns relative priorities for meeting each diversion and instream flow requirement and for maintaining target reservoir storage levels. The solution algorithm computes the flows and storage changes (q_{ij}).

Reservoir/river system operation models. Network flow programming provides considerable flexibility for applying ingenuity in formulating a particular reservoir modeling application. An optimization problem, as formulated earlier, can be solved for each individual time interval in turn, or alternatively a single network flow problem can be solved for all time intervals of the overall period of analysis simultaneously. Convex, piece-wise linear penalty functions can be represented with a q_{ij}, c_{ij} and lower and upper limits (l_{ij} and u_{ij}) on q_{ij} for each linear segment. Successive iterative algorithms are used to handle nonlinearities associated with evaporation and hydropower computations. Generalized reservoir operation models can be designed so that the user simply provides the required input data, with the network flow formulation being somewhat transparent.

A number of reservoir system analysis models which incorporate network flow programming have been reported in the literature. Several such models considered to be representative of the state of the art are discussed next. Two of the models, MODSIM and HEC-PRM, are included in the Model Inventory Appendix.

MODSIM. MODSIM is a generalized river basin network simulation model for hydrologic and water rights analyses of complex water management systems. Water is allocated based on user-specified priorities. The user assigns relative priorities for meeting diversion, instream flow, and storage targets as well as lower and upper bounds on flows and storages. The model computes values for all flows and storages. A network flow programming problem is solved for each individual time interval. Thus, like conventional simulation models, release decisions are not affected by future inflows and future release decisions. Monthly, weekly, or daily time intervals may be used. The model operates in an interactive, menu-driven microcomputer or workstation environment.

MODSIM was developed at Colorado State University and was originally based on modifying and updating the Texas Water Development Board SIMYLD-II model. Various versions of MODSIM are described by Labadie and Shafer (1979), Labadie (1983), Labadie et al. (1984), Frevert et al. (1994), and Labadie et al. (1994). Recent versions of MODSIM include capabilities for analyzing conjunctive use of surface and ground water; water quality considerations; and a graphical user interface. The out-of-kilter algorithm incorporated in earlier versions of MODSIM has been replaced with a more efficient algorithm based on a Lagrangian relaxation strategy. MODSIM has been applied by university researchers and various water management agencies in studies of a number of reservoir/river systems.

HEC-PRM. The Hydrologic Engineering Center (HEC) Prescriptive Reservoir Model (PRM) was originally developed in conjunction with studies of reservoir systems in the Missouri and Columbia River Basins (USACE HEC 1991a, 1991b, 1992, 1993). However, the generalized model can be applied to any reservoir system. HEC-PRM is a network flow programming model designed for prescriptive-oriented applications. Improved network flow computational algorithms have been developed in conjunction with the model. HEC-PRM is used in combination with the HEC-DSS (Data Storage System), which provides input data preparation and output analysis capabilities. Studies to date have used a monthly time interval with historical period of record, or a critical subperiod thereof, hydrology.

Reservoir system operation is driven by user-inputted, convex, cost-based, piece-wise linear penalty functions (dollars versus storage or flow) associated with various purposes, including hydroelectric power, recreation, water supply, navigation, and flood control. The model minimizes a cost-based objective function. Noneconomic components can also be included in the basically economic HEC-PRM objective function. User-specified lower and upper bounds on flows and storages are reflected in the constraint equations. Unlike MODSIM, the HEC-PRM performs the computations simultaneously for all the time intervals. Thus, the HEC-PRM results show a set of reservoir storages and releases which would minimize cost (as defined by the user-inputted penalty functions) for the given inflow sequences, assuming that all future flows are known as release decisions made during each period.

Other network flow programming models.   The Texas Water Development Board (TWDB) began development of a series of models in the late 1960s in conjunction with formulation of the Texas Water Plan. RESOP-II, SIMYLD-II, AL-V, SIM-V, and a number of other models have been developed and modified through various versions. SIMYLD-II, AL-V, and SIM-V incorporate a capacitated network flow formulation solved with the out-of-kilter linear programming algorithm. RESOP-II is also included in this discussion because it can be used in combination with SIMYLD-II.

The Reservoir Operating and Quality Routing Program (RESOP-II) is a conventional simulation model designed for performing a detailed analysis of the firm yield of a single reservoir. A quality routing option adds the capability to route up to three nondegradable constituents through a reservoir and to print a frequency distribution table and a concentration duration plot for the calculated end-of-month quality of the reservoir (Browder 1978).

SIMYLD-II provides the capability for analyzing water storage and transfer within a multireservoir or multibasin system with the objective of meeting a set of specified demands in a given order of priority (Texas Water Development Board 1972). If sufficient water is not available to meet competing demands during a particular time interval, the shortage is assigned to the lowest-priority demand node. SIMYLD-II also determines the firm yield of a single reservoir within a multireservoir system. An iterative procedure is used to adjust the demands at each reservoir of a multireservoir system in order to converge on its maximum firm yield for a given storage capacity, assuming total system operation. Although SIMYLD-II is capable of analyzing multireservoir systems, it is not capable of analyzing a single reservoir as accurately as RESOP-II. Consequently, SIMYLD-II and RESOP-II are used in combination to analyze complex systems.

The Surface Water Resources Allocation Model (AL-V) and Multireservoir Simulation and Optimization Model (SIM-V) simulate and optimize the operation of an interconnected system of reservoirs, hydroelectric power plants, pump canals, pipelines, and river reaches (Martin 1981, 1982, 1983). SIM-V is used to analyze short-term reservoir operations. AL-V is for long-term operations. A system is represented as a network flow programming problem. Hydroelectric benefits, which are complicated by nonlinearity, are incorporated by solving successive network flow problems, where flow bounds and unit costs are modified between successive iterations to reflect first-order changes in hydroelectric power generation with flow release rates and reservoir storage.

Sigvaldason (1976) describes a simulation model developed by Acres Consulting Services to assess alternative operation policies for the 48-reservoir multipurpose water supply, hydropower, and flood control system in the Trent River Basin in Ontario, Canada. The model was originally developed for planning but has also been used for real-time operation. In the model, each reservoir was subdivided into five storage zones, and time-based rule curves were specified. The combined rule curve and storage zone representation is similar to HEC-5. However, the model was formulated in the format of a network flow programming problem. Penalty coefficients were assigned to those variables which represented deviations from ideal conditions. Different operating policies were simulated by altering relative values of these coefficients. The out-of-kilter computational algorithm used to solve the network flow problem is similar to the Texas Water Development Board models. Bridgeman et al. (1988) describe more recent applications of a later version of the network flow model designed to forecast inflows, simulate operations, and postprocess results. Acres International Limited has applied the model to the complex multireservoir Trent River Basin system in both planning and real-time operation modes.

Brendecke et al. (1989) describe the Central Resource Allocation Model (CRAM) developed by WBLA, Inc. for use in preparing a water supply master plan for the city of Boulder, Colorado. The model was used to compute yields which could be achieved with various system operation plans. MODSIM served as the basis for development of CRAM, with various improvements pertinent to the particular application being added to CRAM.

The Water Assignment Simulation Package (WASP) is described by Kuczera and Diment (1988). WASP was developed to analyze the water supply system of the city of Melbourne, Australia, which includes nine reservoirs and a complex conveyance and distribution system, but it is generalized for application to other systems as well. WASP allocates water according to the following criteria in order of decreasing priority: (1) Satisfy all demands, (2) satisfy instream requirements,

(3) minimize spills, (4) ensure that water assignments are consistent with user-defined operating rules, and (5) minimize operating cost. The network programming solution is based on minimizing a weighted penalty function, with a hierarchy of penalties based on the aforementioned priorities.

DWRSIM was developed by the California Department of Water Resources to simulate the combined operation of the State Water Project and Central Valley (Chung et al. 1989). The original DWRSIM was a conventional simulation model, with no mathematical programming, developed based on modifying HEC-3. However, the model was later revised to incorporate the out-of-kilter network flow programming algorithm. The versions of DWRSIM with and without the network flow programming algorithm are used for the same types of analyses and have essentially the same input and output formats. Chung et al. (1989) state that the network flow formulation was incorporated into DWRSIM to enhance capabilities for analyzing consequences of different operational scenarios. The most significant model improvements provided by the network flow algorithm are capabilities to (1) provide a better balance among the reservoirs in the system, (2) assign different relative priorities to different demand points, (3) assign different relative priorities to the different components that make up a demand point, and (4) allocate storage within a reservoir to specific demands.

Generalized Reservoir/River System Analysis Models Based on Object-Oriented Programming

Object-oriented programming has become increasingly popular in recent years. The RSS and CALIDAD modeling systems are flexible packages that combine graphical user interfaces, commercial data management software, and object-oriented simulation models. The models were developed for workstation environments. These object-oriented programming systems provide flexible frameworks for building models for particular reservoir/river system studies.

RSS.   The River Simulation System (RSS) was developed at the Center for Advanced Decision Support for Water and Environmental Systems (CADSWES) under the sponsorship of the Bureau of Reclamation. The generalized computer package is designed to be readily adaptable to any reservoir/river system. As of 1994, the RSS is essentially operational but still in a developmental mode. Improved models are being developed that utilize many of the basic concepts reflected in RSS. One such model is being developed by CADSWES, the Tennessee Valley Authority (TVA), and the Electric Power Research Institute for evaluating multiple-purpose operations of the TVA reservoir system.

The interactive, graphics-based RSS software package is designed to run on workstations using the Unix operating system. The model combines advanced computer graphics and data management technology with river/reservoir system simulation capabilities. Several programs (including S-plus, ARC/INFO, INGRES, HYDAS, and other commercial software) are available within RSS to manage input data and analyze output.

The actual reservoir/river system simulation component of the RSS package is an object-oriented model written in the C and C++ programming languages. River/reservoir systems are represented within the model by node-link components. The user builds a model of a particular system by selecting and combining objects. Preprogrammed instructions for performing computations and data handling functions are associated with each object. For example, the user could select a reservoir object, hydropower object, or diversion object to represent a system component, which results in the model performing certain computations associated with these particular objects. The user defines reservoir system operating rules using "English-like" statements following a specified format. In general, the user can develop a model for a particular system using the preprogrammed objects and functions provided by the RSS. However, the object-oriented program structure also facilitates a programmer altering the software to include additional objects or functions as needed for particular applications.

CALIDAD. CALIDAD is also an object-oriented programming model designed for a workstation environment. The generalized modeling system was recently developed by the Bureau of Reclamation and, as of 1994, was being applied to several river/reservoir systems, including the California Central Valley Project (Boyer 1994). CALIDAD was developed in the C programming language and one of its object-oriented extensions, Objective C. The graphical user interface was developed in C using Motif and the X Intrinsics Libraries.

CALIDAD simulates a river basin system and determines the set of diversions and reservoir releases which best meets the management objectives and institutional constraints. A user-specified water use scenario is supplied for sequences of streamflow inflows. A model for a particular river basin is created by combining a collection of objects representing system features such as stream inflows, reservoirs, municipal or irrigation demand sites, or hydroelectric power plants. CALIDAD provides a palette of precoded objects which can be used for building models for different river basins. Additional objects may be programmed and added to the library as needed. An object developed for a particular application can then be used for other modeling projects as well. Both computational algorithms and data are associated with each object. The physical data for system features such as reservoir storage capacities and streamflows may be entered as object data. Institutional constraints and management objectives, called rules in the model, are also treated as data and entered through a separate rules editor. CALIDAD handles the management and institutional constraints using a heuristic technique, called tabu search, to determine permissible diversion and reservoir releases. If the system is overconstrained, the tabu search selects a release schedule using weighting factors provided by the user.

COMPARISON OF MODELS INCLUDED IN THE MODEL INVENTORY APPENDIX

The preceding review of reservoir/river system analysis models emphasizes the large number and wide variety of available models. The following discussion focuses specifically on seven models, which are listed here and are included in the Model Inventory Appendix.

> HEC-5: Simulation of Flood Control and Conservation Systems—USACE Hydrologic Engineering Center
>
> IRIS: Interactive River System Simulation Model—Cornell University and International Institute for Applied Systems Analysis
>
> TAMUWRAP: Water Rights Analysis Package—Texas A&M University
>
> MODSIM: River Basin Network Simulation Model—Colorado State University
>
> HEC-PRM: Prescriptive Reservoir Model—USACE Hydrologic Engineering Center
>
> CALIDAD: Object-Oriented River Basin Modeling Framework—Bureau of Reclamation
>
> RSS: River Simulation System—Center for Advanced Decision Support for Water and Environmental Systems and U.S. Bureau of Reclamation

The seven models provide a broad range of modeling capabilities. The capabilities of the individual models greatly overlap, but each model has unique features which could make it preferable over the others under appropriate circumstances. Use of two or more of the models in combination might be advantageous in certain situations.

HEC-5 is a very versatile and comprehensive traditional reservoir/river system simulation model which has been used in many studies. IRIS uses interactive computer graphics to enhance communications and interpretation of model results. TAMUWRAP is particularly useful for analyzing systems involving numerous reservoirs and water use requirements controlled by a water rights priority system. CALIDAD and RSS are software packages that employ computer graphics, object-oriented programming, and integration of several support programs. MODSIM is a generalized reservoir/river system simulation/optimization model based on network flow programming. HEC-PRM is a prescriptive optimization model based on network flow programming.

System Representation

All of the models provide capabilities for analyzing operations for water supply, low-flow augmentation, and hydroelectric power. HEC-5 provides particularly flexible options for representing hydroelectric power operating strategies. HEC-5 is the only model which provides capabilities for comprehensive, in-depth analyses of flood control. All of the models allow analysis of systems containing multiple reservoirs and multiple nonreservoir nodes (control points) located on essentially any branching tributary configuration. HEC-5, IRIS, TAMUWRAP, and MODSIM each include significantly different mechanisms for the user to specify operating rules. HEC-PRM does not allow for detailed specification of operating rules, at least not in the same detail as the simulation models. RSS and CALIDAD allow the user to code or write operating rules in a flexible format.

HEC-5 provides particularly flexible options for representing (1) diversion and instream flow targets as a function of reservoir storage and season of the year and (2) reservoir release rules as a function of diversion and instream flow targets, storage, annual season, and streamflow at downstream locations. However, HEC-5, like most conventional simulation models, performs computations proceeding from upstream to downstream and thus is somewhat limited in capabilities for specifying water use priorities. Diversion and instream flow targets are met essentially in upstream-to-downstream order. MODSIM and other simulation models based on network flow programming provide much greater flexibility for assigning priorities for competing water uses and users. TAMUWRAP is a conventional simulation model designed specifically to handle the complexities of a priority-based water rights permit system, involving allocation of water to numerous water users.

The object-oriented modeling approach of RSS and CALIDAD provides the user a set of versatile tools for building a model for the particular reservoir/river system of interest. A library of existing objects can be used to build and modify models. The object-oriented structure of the computer code also facilitates reprogramming the model as needed to create additional objects or otherwise provide flexibility in representing a particular system.

Measures of System Performance

Each of the simulation models has certain unique features for organizing the simulation results in concise measures of hydrologic performance or reliability. For example, whereas iterative runs of the other simulation models are required to compute firm yields, HEC-5 includes an automated search procedure for determining reservoir firm yields in a single run.

HEC-PRM is designed specifically to use an economic objective function formed from convex, cost-based, piece-wise linear penalty functions. The relative priorities reflected in the MODSIM objective function can also be based on economics if the user so chooses. Because the models simply assign dollars as a function of storages and flows using user-inputted penalty functions, the same type of economic analysis, for user-specified operating plans, could be used with any of the simulation models. There is no need to limit economic functions to the convex, piece-wise linear format when used with a conventional simulation model. HEC-5 is the only model that provides capabilities for detailed economic evaluation of flood control plans, including computing expected annual flood damages.

Prescriptive versus Descriptive Orientation

HEC-PRM is a prescriptive optimization model. Although MODSIM can be used in a more prescriptive mode if the user so chooses, it is designed to be essentially a descriptive simulation model, even though it also is built on network flow programming. The other five models are descriptive simulation models. The prescriptive HEC-PRM has the advantage of automatically finding a set of release and storage values which minimize an objective function designed to capture the actual planning objective of economic efficiency. The descriptive simulation models do not automatically find an optimal set of release and storage values but do show the releases and storages which would result from a particular operating plan. The simulation models have the advantage of allowing the user to define much more precisely the operating plan being simulated. The advantages of both types of models can be realized by using the models in combination.

Interpretation and Communication of Results

All of the models provide voluminous output, including (for each time interval) computed storages, releases, diversions, instream flows, and hydroelectric energy generated. The alternative models provide a variety of mechanisms for organizing and summarizing the simulation results in meaningfully concise formats, which include tabular and graphical presentations, summary statistics, storage and flow frequency analyses, reliability measures, and shortage or failure summaries. In recent years, interactive computer graphics has become an important tool in facilitating communications between analysts and water managers. Enhanced visual communication was a major emphasis particularly in the development of IRIS, RSS, and CALIDAD.

REFERENCES

ALLEN, R. B., and S. G. BRIDGEMAN, "Dynamic Programming in Hydropower Scheduling," *Journal of Water Resources Planning and Management,* ASCE, Vol. 112, No.3, July 1986.

BOYER, J. M., "Addressing Central Valley Project Policy Issues Using a General-Purpose Model," in *Water Policy and Management: Solving the Problems,* edited by D. G. Fontane and H. N. Tuvel, American Society of Civil Engineers, 1994.

BRENDECKE, C. M., W. B. DEOREO, E. A. PAYTON, and L. T. ROZAKLIS, "Network Models of Water Rights and System Operations," *Journal of Water Resources Planning and Management,* ASCE, Vol. 115, No. 5, September 1989.

BRIDGEMAN, S. G., D. J. W. NORRIE, H. J. COOK, and B. KITCHEN, "Computerized Decision-Guidance System for Management of the Trent River Multireservoir System," in *Computerized Decision Support Systems for Water Managers,* edited by Labadie, Brazil, Corbu, and Johnson, ASCE, 1988.

BROWDER, L. E., "RESOP-II Reservoir Operating and Quality Routing Program, Program Documentation and User's Manual," UM-20, Texas Department of Water Resources (now renamed Texas Water Development Board), Austin, Texas, 1978.

CHUNG, F. I., M. C. ARCHER, and J. J. DEVRIES, "Network Flow Algorithm Applied to California Aqueduct Simulation," *Journal of Water Resources Planning and Management,* ASCE, Vol. 115, No. 2, March 1989.

CHUNG, F. I., and O. HELWEG, "Modeling the California State Water Project," *Journal of Water Resources Planning and Management,* ASCE, Vol. 111, No. 1, January 1985.

CIESLIK, P. E., and R. F. MCALLISTER, "Missouri River Master Manual Review and Update," in *Water Policy and Management: Solving the Problems,* edited by D. G. Fontane and H. N. Tuvel, American Society of Civil Engineers, 1994.

COLON, R., and G. F. MCMAHON, "BRASS Model: Application to Savannah River System Reservoirs," *Journal of Water Resources Planning and Management,* ASCE, Vol. 113, No. 2, March 1987.

FELDMAN, A. D., "HEC Models for Water Resources System Simulation: Theory and Experience," in *Advances in Hydroscience,* edited by V. T. Chow, vol. 12, Academic Press, 1981.

FREVERT, D. K., J. W. LABADIE, ROGER K. LARSON, and N. L. PARKER, "Integration of Water Rights and Network Flow Modeling in the Upper Snake River Basin," in *Water Policy and Management: Solving the Problems,* edited by D. G. Fontane and H. N. Tuvel, American Society of Civil Engineers, 1994.

GILBERT, K. C., and R. M. SHANE, "TVA Hydro Scheduling Model: Theoretical Aspects," *Journal of Water Resources Planning and Management,* ASCE, Vol. 108, No. 1, March 1982.

GILES, J. E., and W. O. WUNDERLICH, "Weekly Multipurpose Planning Model for TVA Reservoir System," *Journal of Water Resources Planning and Management,* ASCE, Vol. 107, No. 2, October 1981.

HULA, R. L., "Southwestern Division Reservoir Regulation Simulation Model," in *Proceed-*

ings of the National Workshop on Reservoir Systems Operations, edited by Toebes and Sheppard, ASCE, New York, 1981.

JENSEN, P. A., and J. W. BARNES, *Network Flow Programming,* John Wiley and Sons, 1980.

KENNINGTON, J. L., and R. V. HELGASON, *Algorithms for Network Programming,* Wiley-Interscience, 1980.

KUCZERA, G., and G. DIMENT, "General Water Supply System Simulation Model: Wasp," *Journal of Water Resources Planning and Management,* ASCE, Vol. 114, No. 4, July 1988.

LABADIE, J. W., "Drought Contingency Model for Water Control at Corps Reservoirs in the Ohio River Basin: Cumberland Basin Study," Reservoir Control Center, USACE Ohio River Division, Cincinnati, Ohio, March 1983.

LABADIE, J. W., "Dynamic Programming with the Microcomputer," in *Encyclopedia of Microcomputers,* Marcel Dekker, Inc., 1990.

LABADIE, J. W., A. M. PINEDA, and D. A. BODE, "Network Analysis of Raw Supplies Under Complex Water Rights and Exchanges: Documentation for Program MODSIM3," Colorado Water Resources Institute, Fort Collins, Colorado, March 1984.

LABADIE, J. W., D. G. FONTANE, and T. DAI, "Integration of Water Quantity and Quality in River Basin Network Flow Modeling," in *Water Policy and Management: Solving the Problems,* edited by D. G. Fontane and H. N. Tuvel, American Society of Civil Engineers, 1994.

LABADIE, J. W., and J. M. SHAFER, "Water Management Model for Front Range River Basins," Technical Report No. 16, Colorado Water Resources Research Institute, April 1979.

LOUCKS, D. P., J. R. STEDINGER, and D. A. HAITH, *Water Resource Systems Planning and Analysis,* Prentice Hall, 1981.

LOUCKS, D. P., K. A. SALEWICZ, and M. R. TAYLOR, "IRIS: An Interactive River System Simulation Model, General Introduction and Description," Cornell University, Ithaca, New York, and International Institute for Applied Systems Analysis, Laxenburg, Austria, November 1989.

LOUCKS, D. P., K. A. SALEWICZ, and M. R. TAYLOR, "IRIS: An Interactive River System Simulation Model, User's Manual," Cornell University, Ithaca, New York, and International Institute for Applied Systems Analysis, Laxenburg, Austria, January 1990.

LOUCKS, D. P., P. N. FRENCH, and M. R. TAYLOR, "IRAS: Interactive River-Aquifer Simulation, Program Description and Operation," Cornell University and Resource Planning Associates, Ithaca, New York, Draft, January 1994.

MARTIN, Q. W., "Surface Water Resources Allocation Model (AL-V), Program Documentation and User's Manual," UM-35, Texas Department of Water Resources, Austin, Texas, October 1981.

MARTIN, Q. W., "Multivariate Simulation and Optimization Model (SIM-V), Program Documentation and User's Manual," UM-38, Texas Department of Water Resources, Austin, Texas, March 1982.

MARTIN, Q. W., "Optimal Operation of Multiple Reservoir Systems," *Journal of Water Resources Planning and Management,* ASCE, Vol. 109, No. 1, January 1983.

MARTIN, Q. W., "Optimal Daily Operation of Surface Water Systems," *Journal of Water Resources Planning and Management,* ASCE, Vol. 113, No. 4, July 1987.

MAYS, L. W., and Y-K. TUNG, *Hydrosystems Engineering and Management,* McGraw-Hill, 1992.

PALMER, R. N., J. A. SMITH, J. L. COHON, and C. S. REVELLE, "Reservoir Management in Potomac River Basin," *Journal of Water Resources Planning and Management,* ASCE, Vol. 108, No. 1, March 1982.

PALMER, R. N., and K. J. HOLMES, "Operational Guidance During Droughts: Expert System Approach," *Journal of Water Resources Planning and Management,* ASCE, Vol. 114, No. 6, November 1988.

PATENODE, G. A., and K. L. WILSON, "Development and Application of Long Range Study (LRS) Model for Missouri River System," in *Water Policy and Management: Solving the Problems,* edited by D. G. Fontane and H. N. Tuvel, American Society of Civil Engineers, 1994.

RANDALL, D., M. H. HOUCK, and J. R. WRIGHT, "Drought Management of Existing Water Supply System," *Journal of Water Resources Planning and Management,* ASCE, Vol. 116, No. 1, January 1990.

SCHUSTER, R. J., "Colorado River Simulation System, Executive Summary," U.S. Bureau of Reclamation, Denver, Colorado, April 1987.

SHANE, R. M., and K. C. GILBERT, "TVA Hydro Scheduling Model: Practical Aspects," *Journal of Water Resources Planning and Management,* ASCE, Vol. 108, No. 1, March 1982.

SIGVALDASON, O. T., "A Simulation Model for Operating a Multipurpose Multireservoir System," *Water Resources Research,* AGU, Vol. 12, No. 2, April 1976.

SIMONOVIC, S. P., "Reservoir Systems Analysis: Closing Gap between Theory and Practice," *Journal of Water Resources Planning and Management,* ASCE, Vol. 118, No. 3, May 1992.

STRZEPEK, K. M., L. A. GARCIA, and T. M. OVER, "MITSIM 2.1 River Basin Simulation Model, User Manual," Center for Advanced Decision Support for Water and Environmental Systems, University of Colorado, Draft, May 1989.

STRZEPEK, K. M., M. S. ROSENBERG, D. D. GOODMAN, R. L. LENTON, and D. D. MARKS, "User's Manual for MIT River Basin Simulation Model," Report 242, Civil Engineering Department, Massachusetts Institute of Technology, July 1979.

TEXAS WATER DEVELOPMENT BOARD, "Economic Optimization and Simulation Techniques for Management of Regional Water Resource Systems: River Basin Simulation Model SIMYLD-II Program Description," Austin, Texas, July 1972.

U.S. ARMY CORPS OF ENGINEERS, BALTIMORE DISTRICT, "Metropolitan Washington D.C. Area Water Supply Study, Final Report," September 1983.

U.S. ARMY CORPS OF ENGINEERS, HYDROLOGIC ENGINEERING CENTER, "HEC-3 Reservoir System Analysis for Conservation, User's Manual," March 1981.

U.S. ARMY CORPS OF ENGINEERS, HYDROLOGIC ENGINEERING CENTER, "HEC-5 Simulation of Flood Control and Conservation Systems, User's Manual," April 1982.

U.S. ARMY CORPS OF ENGINEERS, HYDROLOGIC ENGINEERING CENTER, "HEC-5 Simulation of Flood Control and Conservation Systems, Exhibit 8, Input Description," January 1989.

U.S. ARMY CORPS OF ENGINEERS, HYDROLOGIC ENGINEERING CENTER, "Columbia River System Model—Phase I," PR-16, October 1991a.

U.S. ARMY CORPS OF ENGINEERS, HYDROLOGIC ENGINEERING CENTER, "Missouri River System Analysis Model—Phase I," PR-15, February 1991b.

U.S. ARMY CORPS OF ENGINEERS, HYDROLOGIC ENGINEERING CENTER, "Optimization of Multiple-Purpose Reservoir System Operations: A Review of Modeling and Analysis Approaches," Research Document No. 34, January 1991c.

U.S. ARMY CORPS OF ENGINEERS, HYDROLOGIC ENGINEERING CENTER, "Missouri River Reservoir System Analysis Model: Phase II," PR-17, January 1992.

U.S. ARMY CORPS OF ENGINEERS, HYDROLOGIC ENGINEERING CENTER, "Columbia River System Model—Phase II," PR-21, December 1993.

U.S. ARMY CORPS OF ENGINEERS, NORTH PACIFIC DIVISION, "User Manual, SSARR Streamflow Synthesis and Reservoir Regulation," Portland, Oregon, August 1987.

U.S. ARMY CORPS OF ENGINEERS, OFFICE OF THE CHIEF OF ENGINEERS, "Engineering and Design, Hydropower," EM 1110-2-1701, 31 December 1985.

U.S. BUREAU OF RECLAMATION, "Inventory of Hydrologic Models," Global Climate Change Response Program, Denver, Colorado, August 1991.

WURBS, R. A., "Reservoir Simulation and Optimization Models," *Journal of Water Resources Planning and Management,* American Society of Civil Engineers, Vol. 119, No. 4, July/August 1993.

WURBS, R. A., D. D. DUNN, and W. B. WALLS, "Water Rights Analysis Program (TAMU-WRAP), Model Description and Users Manual," Technical Report 146, Texas Water Resources Institute, College Station, Texas, March 1993.

WURBS, R. A., G. SANCHEZ-TORRES, and D. D. DUNN, "Reservoir/River System Reliability Considering Water Rights and Water Quality," Technical Report 165, Texas Water Resources Institute, College Station, Texas, March 1994.

WURBS, R. A., M. N. TIBBETS, L. M. CABEZAS, and L. C. ROY, "State-of-the-Art Review and Annotated Bibliography of Systems Analysis Techniques Applied to Reservoir Operation," Technical Report 136, Texas Water Resources Institute, College Station, Texas, June 1985.

WURBS, R. A., and W. B. WALLS, "Water Rights Modeling and Analysis," *Journal of Water Resources Planning and Management,* ASCE, Vol. 115, No. 4, July 1989.

YEH, W. W-G., "Real-Time Reservoir Operation: The Central Valley Project Case Study," in *Proceedings of the National Workshop on Reservoir Operations,* edited by Toebes and Sheppard, ASCE, 1981.

YEH, W. W-G., "Reservoir Management and Operations Models: A State-of-the-Art Review," *Water Resources Research,* AGU, Vol. 21, No. 21, December 1985.

MODEL INVENTORY APPENDIX

DEMAND FORECASTING AND BALANCING SUPPLY WITH DEMAND MODELS (CHAPTER 4)

(IWR-MAIN) Water Use Forecasting System

Contact   Institute for Water Resources
U.S. Army Corps of Engineers
Casey Building
7701 Telegraph Rd.
Alexandria, VA 22310-3868
(703)355-2015

Planning and Management
Consultants, Ltd.
P.O. Box 1316
Carbondale, IL 62903
(618)549-2832

Model Availability   The computer program and documentation can be obtained by contacting the Institute for Water Resources (IWR) or Planning and Management Consultants, Ltd. Future plans include distribution of the model by the American Public Works Association through a users' group.

Documentation   IWR-MAIN Version 5.1 is documented by Davis, Rodrigo, Opitz, Dziegielewski, Baumann, and Boland, "IWR-MAIN Water Use Forecasting System Version 5.1 Users Manual and System Description," IWR Report 88-R-6, June 1988, revised August 1991. A user's guide for the new Version 6.0 is being prepared. Several IWR publications describe various methods incorporated into IWR-MAIN. Planning and Management Consultants, Ltd., periodically offers a training course on application of IWR-MAIN, in coordination with IWR and the American Public Works Association.

Computer Configuration   A compiled version of IWR-MAIN is available for MS-DOS-based microcomputers. The FORTRAN77 program has been compiled and executed on other computer systems as well. IWR-MAIN is an interactive, menu-driven modeling system.

Capabilities   IWR-MAIN is a flexible software package for predicting future municipal and industrial water use. The forecasting system provides a variety of forecasting models, socioeconomic parameter generating procedures, and data management capabilities. A high level of disaggregation of water use categories is provided. Water requirements are estimated separately for residential, commercial/institutional, industrial, and public/unaccounted sectors. Within these major sectors, water use estimates are further disaggregated into individual categories such as metered and sewered residences, commercial establishments, and three-digit SIC manufacturing categories. Average daily water use, winter and summer daily use, and maximum-day summer use are forecasted as a function of explanatory variables which include number of users; number, market value, and type of housing units; employment in commercial and manufacturing industries; water and wastewater fee rates; irrigated acreage; climatic conditions; and water conservation measures.

Experience   Various versions of IWR-MAIN have been applied in a number of studies of various cities located throughout the United States during the past several years.

(WEAP) Water Evaluation and Planning System

Contact   Tellus Institute
Stockholm Environment Institute, Boston Center
11 Arlington Street
Boston, MA 02116-3411
(617)266-5400

Model Availability   The proprietary computer program and documentation can be purchased from the Tellus Institute. The price is $1,500 for private entities and $750 for universities, government agencies, and nonprofit organizations.

Documentation   WEAP Version 93.0 is documented by Tellus Institute, Stockholm

Environment Institute Boston Center, "WEAP, a Computerized Water Evaluation and Planning System, User Guide," April 1993.

Computer Configuration   An executable version of WEAP is available for MS-DOS-based microcomputers. The program is written in BASIC. WEAP is an interactive, menu-driven modeling system.

Capabilities   WEAP provides a systematic framework for accounting for and comparing the water supplies and uses for any particular geographic area. The model provides capabilities for database management, water use forecasting, and analysis of supplies and demands. Alternative water use scenarios and management strategies may be evaluated. Water supply and use data may be displayed in a variety of tables and graphs. Network diagrams are available that show the interconnected relationships between the components of the water demand-supply system. Disaggregated demand forecasts may be performed for municipal, industrial, agricultural, and other types of water use over a long-term planning horizon. Several optional forecasting methods are available. All surface and ground-water supplies can be included in a simulation. Major multiple-purpose reservoirs as well as local water supply reservoirs may be modeled. Withdrawals for water treatment plants, discharges from wastewater treatment plants, return flows, ground-water pumpage, and transmission losses are included in the water accounting system. Supplies and demands are compared at a site-specific level, such as a water treatment or wastewater treatment plant, or at an aggregate level such as a city or county. Streamflow data can be entered for the historical period of record or a critical drought period, or alternatively streamflows can be entered characterizing typical wet, dry, and normal years. The model uses a monthly time interval.

Experience   The model is relatively new. It has been applied in several studies in different regions of the world.

WATER DISTRIBUTION SYSTEM ANALYSIS MODELS (CHAPTER 5)

(KYPIPE2) University of Kentucky Pipe Network Model

Contact   Don J. Wood or William C. Gilbert
Civil Engineering Software Center
212 Anderson Hall
University of Kentucky
Lexington, KY 40506-0046
(606)257-3436 or (606)257-4941

Haestad Methods, Inc.
37 Brookside Road
Waterbury, CT 06708
(203)755-1666
(800)727-6555

Model Availability The proprietary computer program and documentation can be obtained from the Civil Engineering Software Center at the University of Kentucky. The KYPIPE2 simulation program and associated pre- and post-processor programs can be purchased as a package or as individual programs. Various versions of KYPIPE2 are dimensioned for different sizes of pipe networks, and the price varies between versions. A version distributed for educational purposes can handle up to 60 pipes. An extended memory micro-computer version of KYPIPE2 for analyzing large networks of up to 6,000 pipes sells for $700. The price is less for microcomputer versions not requiring extended memory, which are dimensioned for networks not exceeding 1,800 pipes. Haestad methods also sells KYPIPE2 and CYBERNET, which is an AutoCAD-based network model with the KYPIPE2 computational algorithms embedded. The price of CYBERNET ranges from $995 to $7,995 depending on the limits on the size of the pipe network which can be modeled.

Documentation The model is documented by Don J. Wood, "KYPIPE2 User's Manual, Comprehensive Computer Modeling of Pipe Distribution Networks," Civil Engineering Software Center, University of Kentucky, November 1991. On-line help information is also provided by the interactive software. Various universities and firms periodically offer short courses.

Computer Configuration Compiled versions of KYPIPE2 and associated utility programs are available for MS-DOS-based microcomputers. Versions for networks exceeding 1,800 pipes require extended memory. The FORTRAN programs have been compiled and executed on other computer systems as well. KYPIPE2 and the input preparation and output display programs are accessed by the user through an interactive menu structure.

Capabilities The comprehensive pipe network modeling system computes steady-state flows and pressures for specified demands. Optional capabilities are also provided for extended period simulations with storage tank levels varying over time. The model will compute flows for each pipe and the hydraulic grade and pressure at each node for a given set of water demands. Alternatively, capabilities are provided to determine a variety of design, operation, and calibration parameters for specified pressure requirements. Simulations are based on iteratively solving the sets of continuity and energy equations using linearization schemes to handle nonlinear terms and a sparse matrix solution algorithm. Pipe head losses are estimated using either the Hazen-Williams or Darcy-Weisbach equations.

Experience KYPIPE2 and its predecessor KYPIPE have been applied extensively for over a decade by numerous consulting firms, cities, agencies, and universities.

(WADISO) Water Distribution Simulation and Optimization Model

Contact   Waterways Experiment Station   Lewis Publishers, Inc.
U.S. Army Corps of Engineers   2000 Corporate Blvd. N.W.
3909 Halls Ferry Road   Boca Raton, FL 33431
Vicksburg, MS 39180-6199   (800)272-7737
(601)634-2581

Model Availability   The public domain WADISO was developed at the USACE Waterways Experiment Station and is available through the agency. The model has also been published as the book cited below, along with accompanying diskettes, by Lewis Publishers (CRC Press). The book with diskettes can be purchased for $90. The diskettes contain both the FORTRAN source code and executable program.

Documentation   The model is distributed on diskettes accompanying the following book: T. M. Walski, J. Gessler, and J. W. Sjostrom, *Water Distribution Systems: Simulation and Sizing,* Lewis Publishers, 1990. The book is both a text on practical application of water distribution system modeling and a user's manual for WADISO.

Computer Configuration   WADISO is an interactive, menu-driven modeling system which is furnished as executable files for MS-DOS-based microcomputers as well as FORTRAN source code. The interactive structure facilitates development and management of input data files, running of appropriate program modules, and analysis of results.

Capabilities   The WADISO model consists of three major modules or routines: simulation, optimization, and extended period simulation. The simulation routine calculates the flow and pressure distributions in a pipe network for specified demands. The optimization routine determines costs and some pressure distribution for a set of user-specified pipe sizes and changes the sizes for selected pipes within user-specified limits until it finds the most economical arrangement that meets the pressure requirement. The extended period simulation module computes flow and pressure distributions in a pipe network, taking into consideration fluctuating storage tank water levels and varying water use patterns over time. All three routines allow for the presence of pumps, pressure-reducing valves, check valves within the water distribution system, and multiple supply points. Pipe head losses are estimated using the Hazen-Williams equation. There are no limitations to the layout of a system except for the normal requirement of at least one constant head node, such as a tank or reservoir. The optimization routine is intended for sizing of a limited number of pipes, not all the pipes in a large system. Typically, this component of the model is used to size the pipes in an expansion of an existing system or to improve the pressure conditions by cleaning selected pipes.

Experience   WADISO has been applied to a large number of water distribution systems in various parts of the world by the model developers and others.

GROUND WATER MODELS (CHAPTER 6)

(MODFLOW) Modular Three-Dimensional Finite-Difference Ground-Water Flow Model

Contact   Chief Hydrologist
Water Resources Division
U.S. Geological Survey
409 National Center
Reston, VA 22092
(703)648-5215

International Ground Water
   Modeling Center
Colorado School of Mines
Golden, CO 80401-1887
(303)273-3103

Scientific Software Group
P.O. Box 23041
Washington, DC 20026-3041
(703)620-6793

Model Availability   Versions of MODFLOW and associated preprocessor, postprocessor, and other related programs are available from the U.S. Geological Survey, International Ground Water Modeling Center, Scientific Software Group, and others. A nominal handling fee is charged. These entities also provide model users with limited technical support.

Documentation   MODFLOW is documented by M. G. McDonald and A. W. Harbaugh, "A Modular Three-Dimensional Finite-Difference Ground-Water Flow Model," U.S. Geological Survey, 1988. A MODFLOW instructional manual and example data sets are available from the International Ground Water Modeling Center.

Computer Configuration   An executable version of the model is available for MS-DOS-based microcomputers. The FORTRAN77 program has been compiled and executed on various other computer systems as well.

Capabilities MODFLOW simulates two-dimensional areal or cross-sectional and quasi- or fully three-dimensional steady or transient saturated flow in aniso-tropic, heterogeneous, layered aquifer systems. Layers may be simulated as confined, unconfined, or convertible between the two conditions. The model allows for analysis of external influences such as wells, areal recharge, drains, evapotranspiration, and streams. MODFLOW incorporates a block-centered, finite difference approach. The finite difference equations are solved by either the strongly implicit procedure or the slice-successive overrelaxation procedure.

Experience MODFLOW is probably the most widely used of all the available ground-water flow models. The model has been applied in numerous studies throughout the United States and in other countries as well.

(PLASM) Prickett-Lonnquist Aquifer Simulation Model

Contact Thomas A. Prickett & Associates
6 GH Baker Drive
Urbana, IL 61801
(217)384-0615

International Ground Water Modeling Center
Colorado School of Mines
Golden, CO 80401-1887
(303)273-3103

Scientific Software Group
P.O. Box 23041
Washington, DC 20026-3041
(703)620-6793

Model Availability PLASM was originally published by the Illinois State Water Survey. Updated versions of the model, associated utility programs, and documentation are available from the International Ground Water Modeling Center, Thomas A. Prickett & Associates, Scientific Software Group, and others. A nominal handling fee is charged. These entities also provide model users with limited technical support.

Documentation The documentation distributed with the updated computer model includes installation instructions and the original report: T. A. Prickett and C. G. Lonnquist, "Selected Digital Computer Techniques for Groundwater Resource Evaluation," Illinois State Water Survey, Bulletin 55, 1971.

Computer Configuration An executable version of the model is available for MS-DOS-based microcomputers. The FORTRAN program has been compiled and executed on various other computer systems as well.

Capabilities PLASM simulates two-dimensional unsteady flow in heterogeneous anisotropic aquifers under water table, nonleaky, and leaky artesian conditions. The model allows representation of time varying pumpage from wells, natural or artificial recharge rates, the relationships of water exchange between surface waters and the ground-water reservoir, the process of ground-water evapotranspiration, and the mechanism of converting from artesian to water table conditions. PLASM incorporates an iterative alternating direction, implicit finite difference solution of the equations of ground-water flow.

Experience PLASM was one of the first readily available, well-documented ground-water flow models. It continues to be used widely. Various updated microcomputer versions of the model have been developed and applied. PLASM has also been incorporated as the flow component of the RANDOM WALK solute transport model.

(RANDOM WALK) Random Walk Solute Transport Model

Contact Thomas A. Prickett & Associates   International Ground Water
6 GH Baker Drive   Modeling Center
Urbana, IL 61801   Colorado School of Mines
(217)384-0615   Golden, CO 80401-1887
                                                       (303)273-3103

Model Availability RANDOM WALK was originally published by the Illinois State Water Survey. The model, with documentation, is also available through the International Ground Water Modeling Center and Thomas A. Prickett & Associates. A nominal handing fee is charged.

Documentation The model is documented by T. A. Prickett, T. G. Naymik, and C. G. Lonnquist, "A Random-Walk Solute Transport Model for Selected Ground-water Quality Evaluations," Illinois State Water Survey, Bulletin 65, 1981.

Computer Configuration An executable version of the model is available for MS-DOS-based microcomputers. The FORTRAN program has been compiled and executed on various other computer systems as well.

Capabilities PLASM, described earlier, is incorporated into the RANDOM WALK model to perform the flow computations. Thus, RANDOM WALK provides the same capabilities as PLASM for simulating nonsteady or steady, one- or two-dimensional flow. In addition, contaminant transport is simulated using discrete parcel random walk techniques. Solute transport is based on a particle-in-a-cell technique for advective mechanisms and a random walk technique for dispersion mechanisms. The effects of convection, dispersion, and chemical reactions are included. The solute transport model simulates continuous and slug contaminant source areas of various shapes, contaminant sinks such as wells and streams, vertically averaged saltwater fronts, and contaminant leakage from overlying source beds.

Experience   The RANDOM WALK model has been applied widely in numerous studies.

(MOC) Method of Characteristics Two-Dimensional Solute Transport Model

Contact   Chief Hydrologist
Water Resources Division
U.S. Geological Survey
409 National Center
Reston, VA 22092
(703)648-5215

International Ground Water
Modeling Center
Colorado School of Mines
Golden, CO 80401-1887
(303)273-3103

Scientific Software Group
P.O. Box 23041
Washington, DC 20026-3041
(703)620-6793

Model Availability   Updated versions of MOC and associated utility programs and documentation are available from the U.S. Geological Survey, International Ground Water Modeling Center, Scientific Software Group, and others. A nominal handling fee is charged. These entities also provide model users with limited technical support.

Documentation   The MOC model is documented by L. F. Konikow and J. D. Bredehoeft, "Computer Program of Two-Dimensional Solute Transport and Dispersion in Ground Water," Techniques of Water Resources Investigations, Book 7, Chapter C2, U.S. Geological Survey, 1978.

Computer Configuration   An executable version of the model is available for MS-DOS-based microcomputers. The FORTRAN program has been compiled and executed on various other computer systems as well.

Capabilities   MOC is a two-dimensional, transient, saturated condition, solute transport model. MOC allows modeling of heterogeneous and anisotropic (confined) aquifers. The model determines changes in contaminant concentrations caused by convective transport, hydrodynamic dispersion, mixing or dilution from recharge, and chemical reactions. The chemical reactions include first-order irreversible rate reactions (such as radioactive decay); reversible equilibrium-controlled sorption with linear, Freundlich, or Langmuir isotherms; and reversible equilibrium-controlled ion exchange for monovalent or divalent ions. The model assumes that fluid density variations, viscosity changes, and temperature gradients do not affect the velocity changes, and temperature gradients do not affect the velocity distribution. MOC solves the ground-water flow equation and the nonconservative solute transport equation in a stepwise uncoupled fashion. The alternating direction implicit method or the strongly implicit procedure are optionally used to solve the finite difference approximation of the flow equation. The MOC model uses the method of characteristics (MOC) to solve the solute transport equation.

Experience   MOC has been used widely in numerous studies.

(WHPA) Wellhead Protection Area Model

Contact   Center for Subsurface Modeling
Support
Kerr Environmental Research
Laboratory
U.S. Environmental Protection
Agency
Ada, OK 74820
(405)436-8500

International Ground Water
Modeling Center
Colorado School of Mines
Golden, CO 80401-1887
(303)273-3103

Model Availability   WHPA was developed by the EPA and is available from both the Robert S. Kerr Environmental Research Laboratory and the IGWMC.

Documentation   A comprehensive user's manual is distributed along with the computer program.

Computer Configuration   An executable version of the model is available for MS-DOS-based microcomputers. The FORTRAN program has been compiled and executed on various other computer systems as well.

Capabilities   WHPA was originally designed to assist federal, state, and local technical personnel in the delineation of Wellhead Protection Areas as defined by the 1986 Amendments to the Safe Drinking Water Act, but the model can be applied to many different problems associated with wells. WHPA requires relatively few parameters, which makes it useful as a screening tool in delineating capture zones or contaminant fronts. WHPA consists of four independent computational modules. Three of the modules contain semianalytical solutions for delineating capture zones assuming homogeneous aquifers with two-dimensional, steady-state ground-water flow in an areal plane. Multiple pump and injection wells may be present. Barrier or stream boundary conditions that exist over the entire aquifer depth may be simulated. One of these three modules provides an uncertainty analysis of the computed results based on Monte Carlo techniques. The fourth module is a general particle tracking routine that may be used as a postprocessor for two-dimensional numerical ground-water flow models. Because a numerical flow model, such as MOD-FLOW, is used to determine hydraulic heads, the hydrogeologic scenarios that may be investigated with this module are limited only by the capabilities of the numerical model. WHPA output is presented as a plot of the capture zone and particle paths. When simulating multiple wells, the respective capture zones and particle paths are shown in different colors. Plots of the results of multiple simulations can be overlayed, one on top of the other, for comparative analysis. A hard copy of the plot as well as a tabulation of the data can be printed using most standard printers or plotters. Plot files also may be transported as input to ARC/INFO or other GIS software.

Experience WHPA has been applied widely in recent years.

(SUTRA) Saturated-Unsaturated Transport Model

Contact Chief Hydrologist
Water Resources Division
U.S. Geological Survey
409 National Center
Reston, VA 22092
(703)648-5215

International Ground Water
Modeling Center
Colorado School of Mines
Golden, CO 80401-1887
(303)273-3103

Scientific Software Group
P.O. Box 23041
Washington, DC 20026-3041
(703)620-6793

Model Availability Updated versions of SUTRA and associated utility programs and documentation are available from the U.S. Geological Survey, International Ground Water Modeling Center, Scientific Software Group, and others. A nominal handling fee is charged.

Documentation   SUTRA is documented by C. I. Voss, "A Finite-Element Simulation Model for Saturated-Unsaturated, Fluid-Density-Dependent Ground-Water Flow with Energy Transport or Chemically-Reactive Single-Species Solute Transport," U.S. Geological Survey, 1984.

Computer Configuration   An executable version of the model is available for MS-DOS-based microcomputers. The FORTRAN program has been compiled and executed on various other computer systems as well.

Capabilities   SUTRA simulates transient two-dimensional fluid flow and solute and energy transport in a subsurface environment. Two interdependent processes are simulated: (1) fluid density-dependent saturated or unsaturated flow, and either (2a) solute transport, in which the solute may be subject to equilibrium adsorption on the porous matrix and both first-order and zero-order production or decay, or (2b) thermal energy transport in the ground water and solid matrix of the aquifer. The model output includes fluid pressures and either solute concentrations or temperatures, as a function of time and location in the subsurface system. Capabilities are provided for areal and cross-sectional modeling of saturated flow systems and for cross-sectional modeling of the unsaturated zone flow. Boundary conditions, sources, and sinks may be time dependent. SUTRA employs a two-dimensional hybrid finite element and integrated finite difference solution of the governing equations.

Experience   SUTRA has been applied in a number of studies.

(SWIFT-II) Sandia Waste Isolation, Flow, and Transport Model

Contact   GeoTrans, Inc.   Scientific Software Group
            46050 Manekin Plaza, Suite 100   P.O. Box 23041
            Sterling, VA 20166   Washington, DC 20026-3041
            (703)444-7000   (703)620-6793

Model Availability   SWIFT-II was developed by the Sandia National Laboratories for the U.S. Nuclear Regulatory Commission. Updated versions of the model and documentation can be purchased from GeoTrans or the Scientific Software Group. Geotrans also provides technical support for users of the model.

Documentation   SWIFT-II is documented by M. Reeves, D. S. Ward, N. D. Johns, and R. M. Cranwell, "Theory and Implementation for Swift II, The Sandia Waste-Isolation Flow and Transport Model for Fractured Media, Release 4.84," NUREG/CR-3328, SAND83-1159, Sandia National Laboratories,

1986; M. Reeves, D. S. Ward, N. D. Johns, and R. M. Cranwell, "Data Input Guide for Swift II, The Sandia Waste-Isolation Flow and Transport Model for Fractured Media, Release 4.84," NUREG/CR-3162, SAND83-0242, Sandia National Laboratories, 1986; and M. Reeves, D. S. Ward, P. A. Davis, and E. J. Bonano, "Swift II Self-Teaching Curriculum: Illustrated Problems for the Sandia Waste-Isolation Flow and Transport Model for Fractured Media," NUREG/CR-3925, SAND84-1586, Sandia National Laboratories, 1986.

Computer Configuration   An executable version of the model is available for MS-DOS-based microcomputers. The FORTRAN77 program has been compiled and executed on various other computer systems as well.

Capabilities   SWIFT-II is a transient, three-dimensional model, applicable to geologic media which may be fractured, that solves coupled equations for flow and transport. The processes considered are (1) fluid flow, (2) heat transport, (3) dominant-species (brine) miscible displacement, and (4) trace-species (radionuclides) miscible displacement. The first three processes are coupled via fluid density and viscosity. Together they provide the velocity field required in the third and fourth processes. Both dual porosity and discrete fracture conceptualizations may be considered for the fractured zones. SWIFT-II was originally developed for use in the analysis of deep geologic nuclear waste disposal facilities. However, the generalized model is equally applicable to other problem areas, such as injection of industrial wastes into saline aquifers; heat storage in aquifers; *in situ* solution mining; migration of contaminants from landfills; disposal of municipal wastes; saltwater intrusion in coastal regions; and brine disposal from petroleum-storage facilities. A variety of options are provided to facilitate various uses of the model.

Experience   SWIFT-II has been applied in a number of studies.

WATERSHED RUNOFF MODELS (CHAPTER 7)

(HEC-1) Flood Hydrograph Package

Contact   Hydrologic Engineering Center
U.S. Army Corps of Engineers
609 Second Street
Davis, CA 95616
(916)756-1104

Model Availability Hydrologic Engineering Center (HEC) programs, including HEC-1, are available through the National Technical Information Service (NTIS) and various other distributors. Federal agencies can obtain programs directly from the HEC. A HEC computer program catalog and list of distributors are available by contacting the HEC. The NTIS and private distributors charge a nominal handling fee. The programs and documentation can be copied without restriction.

Documentation An HEC-1 user's manual and various other publications are available from the HEC, including a programmer's manual, training documents, and reports and papers documenting specific studies. A free publications catalog is available from the HEC upon request. HEC publications can be ordered directly from the HEC using the price list and order form provided in the catalog. Charges are waived for federal agencies. HEC-1 short courses are taught by the HEC and several universities.

Computer Configuration A compiled version of HEC-1 is available for MS-DOS-based microcomputers. The FORTRAN77 program has been compiled and executed on a variety of other computer systems as well. The HEC-1 microcomputer software package includes several utility programs, linked by a menu structure, which facilitate developing input files and analyzing output. The COED editor and HEC-DSS are often used with HEC-1.

Capabilities HEC-1 models the watershed processes that convert rainfall and/or snowmelt to streamflow, which includes manipulating precipitation data, performing subwatershed precipitation-runoff computations, streamflow routing, and hydrograph combining. HEC-1 is designed for analyzing single precipitation events rather than long-term continuous modeling. Precipitation volumes are converted to runoff volumes using one of the following options: SCS, initial and uniform, exponential, Holtan, or Green and Ampt. Runoff hydrographs are developed using either the unit hydrograph or kinematic wave approaches. Synthetic unit hydrograph options include SCS, Snyder, and Clark. Optional streamflow routing methods include modified Puls, Muskingum, Muskingum-Cunge, working R&D, average lag, and kinematic. HEC-1 also includes several optional modeling capabilities involving parameter calibration, multiplan-multiflood analysis, dam safety analysis, economic flood damage analysis, and flood control system optimization. A version of HEC-1 is also available which facilitates data handling during real-time flood forecasting.

Experience HEC-1 has been applied extensively by the USACE, other agencies, consulting firms, cities, and universities for numerous flood-plain management and other types of studies.

(TR-20) Computer Program for Project Hydrology

Contact   Engineering Division
Soil Conservation Service
US Department of Agriculture
Washington, DC 20013-2890

National Technical Information Service
U.S. Department of Commerce
5285 Port Royal Road
Springfield, VA 22161
(703)487-4600

Model Availability   The computer program and accompanying documentation are available from the National Technical Information Service (NTIS) for a nominal handling fee. Federal agencies can obtain TR-20 directly from the Soil Conservation Service (SCS). Technical assistance is provided by the office of the SCS State Conservation Engineer in each state. The address and telephone number of the SCS State Conservation Engineer for a particular state can be obtained by contacting either NTIS or local SCS offices. The computer program and documentation can be copied without restriction.

Documentation   The model is documented by "Technical Release (TR) 20, Computer Program for Project Formulation, Hydrology," Soil Conservation Service, May 1982 (NTIS number PB83-223768).

Computer Configuration   A compiled version of TR-20 is available for MS-DOS-based microcomputers. The FORTRAN program has been compiled and executed on a variety of other computer systems as well. An interactive input program is provided with TR-20 to assist the user in creating input files.

Capabilities   TR-20 is a single-event watershed model. A rainfall hyetograph is provided as input. The model computes the runoff hydrograph for each subwatershed, routes the hydrographs through reservoirs and stream reaches, and combines hydrographs. The SCS rainfall-runoff relationship (curve number method) and SCS curvilinear dimensionless unit hydrograph are used to model the runoff response of a watershed to a rainfall event. Hydrographs are routed through stream reaches using the attenuation-kinematic routing method. Modified Puls routing is used for reservoirs.

Experience   TR-20 has been applied routinely since 1965 in studies performed by the various offices of the Soil Conservation Service. The model is used widely outside the SCS as well.

A&M Watershed Model

Contact   Dr. Wesley P. James
            Civil Engineering Department
            Texas A&M University
            College Station, TX 77843
            (409)845-4550

Model Availability   The model can be obtained, for a nominal handling fee, by contacting Dr. James at Texas A&M University. The computer program and documentation can be copied without restriction.

Documentation   A user's manual accompanies the model. Various features of the model are discussed by several journal papers cited in Chapter 7.

Computer Configuration   An executable version of the model is available for MS-DOS-based microcomputers. The program is coded in BASIC. The interactive program prompts the user for input data and provides help options.

Capabilities   The A&M Watershed Model simulates a flood event caused by a rainstorm. The model can be used to develop synthetic design storms, or historical gauged rainfall can be provided as input for generating hydrographs for planning and design studies. Alternatively, weather radar and/or gauged rainfall can be used for real-time streamflow forecasting. The model includes the following computational methods: SCS curve number and Green and Ampt loss rate options; two-parameter gamma function unit hydrograph, which can be adjusted by urbanization peaking factors; hydrologic and hydraulic stream routing options; hydrologic reservoir routing; and standard step method water surface profile computations. In addition to the basic rainfall-runoff and streamflow modeling, several optional capabilities are provided for design and analysis of storm sewers, culverts, detention basins, and sedimentation basins. The modeling package also has an option for performing frequency analyses of inputted annual peak discharges using the log-Pearson type III or Gumbel probability distributions.

Experience   The A&M Watershed model is used by a number of consulting firms, cities, and other entities. Many of the users have completed a continuing education short course on the model held at Texas A&M University annually for a number of years.

(SSARR) Streamflow Synthesis and Reservoir Regulation

Contact   North Pacific Division   HOMS National Reference Center
            US Army Corps of Engineers   National Weather Service, NOAA
            P.O. Box 2870   1325 East-West Highway
            Portland, OR 97208-2870   Silver Spring, MD 20910
            (503)326-3758   (301)713-0006

Model Availability   SSARR is available, for a nominal handling fee, by contacting the North Pacific Division. The model is also distributed through the Hydrological Operational Multipurpose System (HOMS) of the World Meteorological Organization. The computer program and documentation can be copied without restriction.

Documentation   The North Pacific Division distributes a user's manual and related documentation with the model.

Computer Configuration   An executable version of the SSARR is available for MS-DOS-based microcomputers. The FORTRAN77 program has been compiled and executed on various other computer systems as well.

Capabilities   SSARR consists of three basic components: (1) a watershed model for synthesizing runoff from rainfall and snowmelt, (2) a streamflow routing model, and (3) a reservoir regulation model. SSARR is a continuous watershed model designed for large river basins. Streamflows are synthesized from rainfall and snowmelt runoff. Rainfall data are provided as input. Snowmelt can be computed based on inputted precipitation depth, elevation, air and dewpoint temperatures, albedo, radiation, and wind speed. Snowmelt options include the temperature index method or the energy budget method. Application of the model begins with a subdivision of the river basin into hydrologically homogeneous subwatersheds. For each subwatershed, the model computes baseflow, subsurface or interflow, and surface runoff. Each flow component is delayed according to different processes, and all are then combined to produce the total subwatershed outflow hydrograph. The subwatershed outflow hydrographs are routed through stream reaches and reservoirs and combined with outflow hydrographs from other subwatersheds to obtain streamflow hydrographs at pertinent locations in the river system.

Experience   Various versions of SSARR date back to 1956. The North Pacific Division initially applied the SSARR to operational flow forecasting and river management activities in the Columbia River Basin. Numerous other river systems in the United States and other countries also have been modeled by various entities using SSARR.

(SWMM) Stormwater Management Model

Contact   Center for Exposure Assessment Modeling
Environmental Research Laboratory
U.S. Environmental Protection Agency
960 College Station Road
Athens, GA 30613-0801
(706)546-3549

Model Availability   Environmental Protection Agency computer programs, including SWWM, are available through the Center for Exposure Assessment Modeling (CEAM) at no charge. Programs can be obtained either through the CEAM Electronic Bulletin Board System or by regular mail on diskette or magnetic tape.

Documentation   The following user's manuals can be obtained on diskette from the CEAM, or paper copies can be ordered through the National Technical Information Service (NTIS): W. A. Huber and R. E. Dickinson, "Storm Water Management Model, Version 4: User's Manual," EPA/600/3-88/001a, EPA Environmental Research Laboratory, August 1988 (NTIS PB88-236-641); and L. A. Roesner, J. A. Aldrich, and R. E. Dickinson, "Storm Water Management Model User's Manual Version 4: EXTRAN Addendum," EPA/600/3-88/001b, EPA Environmental Research Laboratory, August 1988 (NTIS PB88-236-658).

Computer Configuration   A compiled version of SWMM is available for MS-DOS-based microcomputers. The FORTRAN77 program has been compiled and executed on a variety of other computer systems as well.

Capabilities   SWMM is a comprehensive model for analysis of quantity and quality problems associated with urban runoff. Both single-event and continuous simulation may be performed for watersheds that have storm sewers, combined sewers, and natural drainage. Flows, stages, and pollutant concentrations are predicted at pertinent locations in the system. The EXTRAN submodel solves the dynamic flow routing (St. Venant) equations for complex systems that may include backwater, looped connections, surcharging, and pressure flow. The total SWMM package simulates the urban hydrologic and quality processes, including rainfall, snowmelt, surface and subsurface runoff, flow through a drainage system including a sewer network, storage, and treatment.

Options are provided for statistical analysis and presentation of the simulation results.

Experience SWMM has been applied widely by agencies, consulting firms, and university researchers throughout the United States and Canada and has been applied in other countries as well. The EPA CEAM coordinates formal user group conferences and communications to facilitate sharing of experiences by the numerous model users.

(HSPF) Hydrologic Simulation Program—Fortran

Contact Center for Exposure Assessment Modeling
Environmental Research Laboratory
U.S. Environmental Protection Agency
960 College Station Road
Athens, GA 30613-0801
(706)546-3549

Model Availability Environmental Protection Agency computer programs, including HSPF, are available through the Center for Exposure Assessment Modeling (CEAM) at no charge. Programs can be obtained either through the CEAM Electronic Bulletin Board System or by regular mail on diskette or magnetic tape.

Documentation A list of reports and papers regarding HSPF is available from the CEAM. The following user's manual can be obtained on diskette from the CEAM, or paper copies can be ordered through the National Technical Information Service (NTIS): R. C. Johanson, J. C. Imhoff, J. L. Kittle, and A. S. Donigian, "Hydrological Simulation Program—Fortran (HSPF): Users Manual for Release 8.0," EPA-600/3-84-066, EPA Environmental Research Laboratory, 1984 (NTIS PB84-224-385).

Computer Configuration A compiled version of HSPF is available for MS-DOS-based microcomputers. CEAM maintains HSPF on both IBM PC-compatible microcomputers and the DEC/VAX with VMS operating system. The FORTRAN77 source code has been compiled and executed on a variety of other computer systems as well.

Capabilities HSPF is a comprehensive package for simulation of watershed hydrology and water quality for both conventional and toxic organic pollutants. HSPF incorporates the watershed scale ARM (Agricultural Runoff Model) and NPS (Non-Point Source) models into a basin-scale analysis that includes pollutant transport and transformation in stream channels. The model uses information such as the time history of rainfall, temperature, and solar radiation; land surface characteristics such as land use patterns and soil properties; and land management practices to simulate the processes that occur in a watershed. Flow rates, sediment loads, and nutrient and pesticide concentrations are predicted for the watershed runoff. The model uses these results, along with input data characterizing the stream network and point source discharges, to simulate instream processes. Model output includes a time history of water quantity and quality at all pertinent locations in the watershed/stream system. HSPF includes a database management system to process the large amounts of simulation input and output data.

Experience   HSPF and the earlier models from which it was developed have been applied in a variety of hydrologic and water quality studies involving pesticide runoff testing, aquatic fate and transport model testing, analysis of agricultural best management practices, and pesticide exposure assessments.

(SWRRB-WQ) Simulator for Water Resources in Rural Basins—Water Quality

Contact   Grassland, Soil, and Water Research Laboratory
Agricultural Research Service
U.S. Department of Agriculture
808 East Blackland Road
Temple, TX 76502
(817)770-6500

Model Availability   The public domain SWRRB-WQ was developed by the USDA Agricultural Research Service and is available through the Research Laboratory at Temple.

Documentation   The model is distributed on diskettes accompanied by the following documentation: J. G. Arnold, J. R. Williams, A. D. Nicks, and N. B. Sammons, "SWRRB—A Basin Scale Model for Soil and Water Resources Management," Texas A&M University Press, College Station, Texas, 1990; and J. G. Arnold, J. R. Williams, R. H. Griggs, and N. B. Sammons, "SWRRBWQ—A Basin Scale Simulation Model for Assessing Impacts on Water Quality, Model Documentation and User Manual," 1992.

Computer Configuration   An executable version of the model, along with an interactive data editor, is available for microcomputers running MS-DOS. The FORTRAN77 source code is also available.

Capabilities SWRRB-WQ is designed to predict the effects of various types of land management practices on water and sediment yields and water quality in ungauged rural basins. The continuous precipitation-runoff model uses a daily time step. Many years of daily flows may be computed for specified precipitation data. Daily precipitation may be either inputted or developed by the model as a Markov process using inputted probabilities. A large basin can be divided into up to 10 subwatersheds. The major processes in the model include surface runoff, percolation, return flow, evapotranspiration, transmission losses, pond and reservoir storage, sedimentation, nitrogen and phosphorus cycling and movement, pesticide fate and movement, and crop growth and management.

Experience SWRRB-WQ has been used by the Agricultural Research Service, Soil Conservation Service, Environmental Protection Agency, and other agencies to assess the effects of land management on off-site water quantity and quality, pollution of coastal bays and estuaries, reservoir sedimentation, and registration of pesticides.

STREAM HYDRAULICS MODELS (CHAPTER 8)

(HEC-2) Water Surface Profiles

Contact Hydrologic Engineering Center
U.S. Army Corps of Engineers
609 Second Street
Davis, CA 95616
(916)756-1104

Model Availability Hydrologic Engineering Center (HEC) programs, including HEC-2, are available through the National Technical Information Service (NTIS) and various other distributors. Federal agencies can obtain programs directly from the HEC. An HEC computer program catalog and list of distributors are available, free of charge, by contacting the HEC. The NTIS and private distributors charge a nominal handling fee. The programs and documentation can be copied without restriction.

Documentation An HEC-2 user's manual and various other publications are available from the HEC, including a programmer's manual, training documents, and reports and papers documenting specific studies. A free publications catalog is available from the HEC upon request. HEC publications can be ordered directly from the HEC using the price list and order form provided in the catalog. Charges are waived for federal agencies. HEC-2 short courses are taught by the HEC and several universities.

Computer Configuration   A compiled version of HEC-2 is available for MS-DOS-based microcomputers. The FORTRAN77 program has been compiled and executed on a variety of other computer systems as well. The HEC-2 micro-computer software package includes several utility programs, linked by a menu structure, which facilitate developing input files and analyzing output.

Capabilities   HEC-2 develops water surface profiles and related hydraulic data (depths, velocities, etc.) for steady, gradually varied flow in natural and human-made channels. Both subcritical and supercritical flow regimes can be modeled. The computational procedure is based on the standard step method solution of the one-dimensional energy equation with frictional energy losses estimated with the Manning equation. Input data include cross-sections, de-scribing channel and flood-plain geometry, and energy loss coefficients. The effects of various obstructions to flow such as bridges, culverts, weirs, and structures in the flood plain may be reflected in the model. HEC-2 provides optional capabilities for evaluating the effects of channel improvements and levees on water surface profiles. The program also includes options designed for application in flood plain management and flood insurance studies to evaluate floodway encroachments and to designate flood hazard zones. An option is provided for use in calibrating the Manning roughness coefficient. Other optional modeling capabilities involve bridge and culvert losses, stream tributaries, ice-covered streams, and split flows.

Experience   HEC-2 has been applied extensively by the USACE, other agencies, consulting firms, cities, and universities for flood-plain management and flood insurance studies, planning and design of flood control projects, and other types of studies.

(WSPRO) Water Surface Profiles

Contact   U.S. Geological Survey
Water Resources Division
12201 Sunrise Valley Drive
Reston, VA 22092

Federal Highway Administration
Office of Research, Development,
and Technology
6300 Georgetown Pike
McLean, VA 22101-2296

McTrans Center for Microcomputers in Transportation
University of Florida
512 Weil Hall
Gainesville, FL 32611-2083
(904)392-0378

Model Availability   WSPRO and supporting documentation are available from the McTrans Center at the above address. A handling fee of $80 for the program and $25 for documentation is charged. The programs and documentation can be copied without restriction. WSPRO was developed by the USGS in cooperation with the FHWA.

Documentation   WSPRO is documented by J. O. Shearman, W. H. Kirby, V. R. Schneider, and H. N. Filippo, "Bridge Watersway Analysis Model," FHWA/RD86-108, Federal Highway Administration, July 1986; and J. O. Shearman, "User's Manual for WSPRO, a Computer Model for Water Surface Profile Computation," Report No. FHWA-IP-89-027, U.S. Geological Survey, 1990.

Computer Configuration   An executable version of WSPRO is available for MS-DOS-based microcomputers. The FORTRAN77 program has been compiled and executed on various other computer systems as well.

Capabilities   WSPRO develops water surface profiles and related hydraulic data (depths, velocities, etc.) for steady, gradually varied flow in natural and human-made channels. Both subcritical and supercritical flow regimes can be modeled. The computational procedure is based on the standard step method solution of the one-dimensional energy equation with frictional energy losses estimated with the Manning equation. Input data include cross-sections, describing channel and flood-plain geometry, and energy loss coefficients. WSPRO was developed primarily to analyze the hydraulics of bridge waterways. The program provides capabilities for simulating flow through bridges and culverts, including multiple-opening structures, and flows over embankments.

Experience   WSPRO has been applied widely for several years, particularly in the hydraulic design and analysis of bridges and culverts.

(FLDWAV) Flood Wave Model

Contact   Dr. D. L. Fread, Director
Hydrologic Research Laboratory, National Weather Service

National Oceanic and Atmospheric Administration
Silver Spring, MD 20910
(301)713-0006

Model Availability   The public domain FLDWAV as well as DWOPER and DAMBRK are available from the National Weather Service at the above address. A nominal handling fee is charged for each program. The programs and documentation can be copied without restriction.

Documentation   The National Weather Service distributes documentation with the models. The handling fee covers both the computer programs on diskette and the printed documentation.

Computer Configuration   Executable versions of FLDWAV, DWOPER, and DAMBRK are available for MS-DOS-based microcomputers. The FORTRAN programs have been compiled and executed on various other computer systems as well.

Capabilities   FLDWAV combines the widely used NWS Dynamic Wave Operational (DWOPER) Model and Dam-Break Flood Forecasting Model (DAMBRK). These are one-dimensional dynamic routing models based on an implicit finite difference solution of the complete St. Venant equations. Discharges, velocities, depths, and water surface elevations are computed as a function of time and distance along the channel. Input data include cross-sectional geometry of the river and flood plain, energy loss coefficients, and inflow hydrographs. FLDWAV provides flexible capabilities for modeling unsteady flow in rivers with branching tributaries, irregular geometry, variable roughness parameters, lateral inflows, flow diversions, off-channel storage, local head losses such as bridge contractions and expansions, lock and dam operations, and wind effects. An automatic parameter calibration option is provided for determining values for roughness coefficients. Data management features facilitate use of the model in a day-to-day forecasting environment. The model is equally applicable to simulating unsteady flows in planning and design studies. FLDWAV also includes the dam-breach simulation capabilities of DAMBRK. Multiple dams located in series on the same stream can be simulated as well as single dams. An inflow hydrograph is routed through a reservoir using either hydrologic storage routing or dynamic routing. Two types of breaching may be simulated. An overtopping failure is simulated as a trapezoidal-shaped opening that grows progressively downward from the dam crest with time. A piping failure is simulated as a rectangular orifice that grows with time and is centered at any specified elevation within the dam.

Experience DWOPER is used routinely by the National Weather Service River Forecast Centers and has also been applied widely outside the National Weather Service. DAMBRK has been applied extensively by various agencies and consulting firms in conducting dam safety studies. Although the combined FLDWAV was intended to replace DWOPER and DAMBRK, all three models continue to be used widely.

(UNET) One-Dimensional Unsteady Flow through a Full Network of Open Channels

Contact   Hydrologic Engineering Center   Robert L. Barkau, Ph.D., P.E.
U.S. Army Corps of Engineers   5858 Westcliffe Drive
609 Second Street   St. Louis, MO 63129
Davis, CA 95616   (314)846-5152
(916)756-1104

Model Availability UNET is a proprietary program developed by Dr. R. L. Barkau. A HEC version of the model is distributed by the Hydrologic Engineering Center (HEC), under an agreement with Dr. Barkau, following standard HEC procedures. A list of program distributors (vendors) is available from the HEC. Computer programs not available from private vendors can be obtained directly from the HEC. A nominal handling fee is charged for HEC programs, but the programs and documentation can be copied without restriction.

Documentation The following documentation is available from the Hydrologic Engineering Center: R. L. Barkau, "UNET One-Dimensional Unsteady Flow through a Full Network of Open Channels, User's Manual," CPD-66, USACE Hydrologic Engineering Center, May 1993.

Computer Configuration An executable version of UNET is available for MS-DOS-based microcomputers. The system of four FORTRAN77 programs has been compiled and executed on various other computer systems as well. The HEC-DSS (Data Storage System) is often used with UNET to manage data and link with other models.

Capabilities UNET is a dynamic routing model based on a four-point implicit finite difference solution of the St. Venant equations. Unsteady flow can be simulated for complex networks of open channels. Dendritic tributary configurations, split flow around islands, and closed loops, such as a canal connecting tributaries, can be included in the network being modeled. Various types of external and internal boundary conditions can be incorporated in the simulation, including flow and stage hydrographs, rating curves, gated and ungated spillways, pump stations, bridges, culverts, and levee systems. Channel geometry data can be inputted in HEC-2 cross-section format.

Experience UNET has been used in a number of studies both within and outside the Corps of Engineers.

(FESWMS-2DH) Finite Element Surface-Water Modeling System: Two-Dimensional Flow in a Horizontal Plane

Contact U.S. Geological Survey
Water Resources Division
12201 Sunrise Valley Drive
Reston, VA 22092

Federal Highway Administration
Office of Research, Development,
and Technology
6300 Georgetown Pike
McLean, VA 22101-2296

McTrans Center for Microcomputers in Transportation
University of Florida
512 Weil Hall
Gainesville, FL 32611-2083
(904)392-0378

Model Availability FESWMS-2DH and associated documentation are available from the McTrans Center at the above address. A nominal handling fee is charged. The programs and documentation can be copied without restriction.

Computer Configuration An executable version of FESWMS-2DH is available for MS-DOS-based microcomputers. The FORTRAN77 program has been compiled and executed on various other computer systems as well.

Documentation FESWMS-2DH is documented by "FESWMS-2DH, Finite Element Surface-Water System: Two-Dimensional Flow in a Horizontal Plane, User's Manual," Publication No. FHWA-RD-88-177, Federal Highway Administration, Office of Research, Development, and Technology, McLean, Virginia, April 1989.

Capabilities FESWMS-2DH was developed for the Federal Highway Administration by the U.S. Geological Survey to improve capabilities to model complex flow conditions at highway bridges. The generalized model is applicable to other two-dimensional steady or unsteady flow modeling problems as well. FESWMS-2DH is a modular set of programs which includes DINMOD, the data input module; FLOMOD, the depth-averaged flow simulation module; and ANOMOD, the analysis of output module. DINMOD checks the input data for errors, generates plots of the finite element network and ground surface contours, and arranges the network data in the appropriate format. FLOMOD solves the vertically integrated conservation of momentum equations and the conservation of mass (continuity) equation to obtain depth-averaged velocities and water depth at points in a finite element network. ANOMOD generates plots and reports of computed values that simplify interpretation of simulation results. The model is capable of simulating flow through single or multiple bridge openings as normal flow, pressure flow, weir flow, or culvert flow.

Experience   FESWMS-2DH has been used in a number of studies to analyze the impacts of bridges on flows and in other types of studies as well.

(HEC-6) Scour and Deposition in Rivers and Reservoirs

Contact   Hydrologic Engineering Center
U.S. Army Corps of Engineers
609 Second Street
Davis, CA 95616
(916)756-1104

Model Availability   Hydrologic Engineering Center (HEC) programs, including HEC-6, are available through the National Technical Information Service (NTIS) and various other distributors. Federal agencies can obtain programs directly from the HEC. A HEC computer program catalog and list of distributors are available, free of charge, by contacting the HEC. The NTIS and private distributors charge a nominal handling fee. The programs and documentation can be copied without restriction.

Documentation   HEC-6 is documented by the following manual and other publications available from the HEC: "HEC-6 Scour and Deposition in Rivers and Reservoirs, User's Manual," CPD-6, USACE Hydrologic Engineering Center, August 1993.

Computer Configuration   A compiled version of HEC-6 is available for MS-DOS-based microcomputers. The FORTRAN77 program has been compiled and executed on a variety of other computer systems as well.

Capabilities HEC-6 is a one-dimensional sediment transport model designed to develop water surface and sediment bed surface profiles by computing the interaction between sediment material in the streambed and the flowing water-sediment mixture. The model simulates the capability of a stream system to transport bed and suspended load, given the sediment yield from upstream sources. The total sediment load is computed for each cross-section along with the trap efficiencies for clays, silts, and sands. The change in bed elevation, water surface elevation, and thalweg elevation are also computed for each cross-section. HEC-6 does not simulate bank erosion or lateral migration. The model is oriented toward analyzing long-term river and reservoir behavior rather than single short-term flood events. Flow computations are based on a standard step method solution of the steady-state one-dimensional energy equation. The dynamic interactions between bed material composition and transport are based on classical concepts formulated by Einstein. Several user-option alternative sediment transport functions are incorporated in the model.

Experience Versions of HEC-6 date back to the early 1970s. The USACE Waterways Experiment Station, Hydrologic Engineering Center, USACE district offices, and non-Corps entities have applied the model in many studies of various river systems.

RIVER AND RESERVOIR WATER QUALITY MODELS (CHAPTER 9)

(QUAL2E) Enhanced Stream Water Quality Model

Contact Center for Exposure Assessment Modeling
Environmental Research Laboratory
U.S. Environmental Protection Agency
960 College Station Road
Athens, GA 30613-0801
(706)546-3549

Model Availability Several versions of the model have been developed by various entities. The current Environmental Protection Agency release of QUAL2E and QUAL2E-UNCAS can be obtained by contacting the Center for Exposure Assessment Modeling (CEAM). QUAL2E can be obtained, at no charge, either through the CEAM Electronic Bulletin Board System or by regular mail on diskette or magnetic tape.

Documentation   The following documentation can be obtained on diskette from the
CEAM or on paper from the National Technical Information Service (NTIS):
L. C. Brown and T. O. Barnwell, "The Enhanced Stream Water Quality Models
QUAL2E and QUAL2E-UNCAS: Documentation and User Manual,"
EPA/600/3-87/007, EPA Environmental Research Laboratory, May 1987
(NTIS PB87-202-156).

Computer Configuration   An executable version of the model is available for
MS-DOS-based microcomputers. The FORTRAN77 programs have been
compiled and executed on various other computer systems as well.

Capabilities   QUAL2E is a steady-state, one-dimensional model for simulating
pollutant transport and transformation in well-mixed branching streams and
lakes. Up to 15 user-selected water quality constituents in any combination
can be simulated, including dissolved oxygen, biochemical oxygen demand,
temperature, algae as chlorophyll α, organic nitrogen as N, ammonia as N,
nitrite as N, organic phosphorus as P, dissolved phosphorus as P, coliform
bacteria, an arbitrary conservative constituent, and three conservative constituents. A typical application of the model is to study the impacts of waste loads
on stream water quality. QUAL2E also has an option for determining flow
augmentation required to meet any prespecified dissolved oxygen level. The
model can also be used to analyze the effects on water quality, primarily
dissolved oxygen and temperature, caused by diurnal variations in meteorological data. Diurnal dissolved oxygen variations caused by algal growth and
respiration can be examined. QUAL2E-UNCAS is an enhanced version of
QUAL2E which provides capabilities for uncertainty analyses, including sensitivity analysis, first-order error analysis, and Monte Carlo simulations.

Experience   QUAL2E is used widely for waste load allocations, discharge permit
evaluations, and other studies throughout the United States and in other
countries as well. Versions of the model have been developed by various
entities dating back to the late 1960s.

(WASP) Water Quality Analysis Simulation Program

Contact   Center for Exposure Assessment Modeling
Environmental Research Laboratory
U.S. Environmental Protection Agency
Athens, GA 30613
(404)546-3560

Model Availability   The current release of WASP, with documentation on diskette,
can be obtained by contacting the EPA Center for Exposure Assessment

Modeling (CEAM). EPA models, including WASP, are available at no charge, either through the CEAM Electronic Bulletin Board System or by regular mail on diskette or magnetic tape.

Documentation   WASP version 4 (WASP4) is being replaced by version 5 (WASP5). The two versions of the model are documented by R. B. Ambrose, T. A. Wool, and J. P. Connolly, "WASP4, A Hydrodynamic and Water Quality Model—Model Theory, User's Manual, and Programmer's Guide," EPA/600/3-87/039, EPA Environmental Research Laboratory, January 1988; and R. B. Ambrose, T. A. Wool, J. L. Martin, J. P. Connolly, and R. W. Schanz, "WASP5.x, A Hydrodynamic and Water Quality Model—Model Theory, User's Manual, and Programmers's Guide," EPA Environmental Research Laboratory, Draft, 1993.

Computer Configuration   An executable version of the model is available for MS-DOS-based microcomputers. The FORTRAN77 programs have been compiled and executed on various other computer systems as well.

Capabilities   WASP is a compartment modeling framework for simulating contaminant fate and transport in rivers, reservoirs, estuaries, and coastal waters. WASP can be applied in one, two, or three dimensions. The WASP modeling system consists of two stand-alone computer programs, DYNHYD and WASP, that can be run in conjunction or separately. The unsteady flow hydrodynamic program DYNHYD simulates the movement of water, and the water quality program WASP simulates the movement and interaction of pollutants within the water. A variety of water quality problems can be analyzed with the selection of appropriate kinetic subroutines, which may be either selected from a library or written by the user. WASP4 includes two submodels, called EUTRO and TOXI, for simulating two major classes of water quality problems: (1) conventional pollution involving dissolved oxygen, biochemical oxygen demand, nutrients, and eutrophication, and (2) toxic pollution involving organic chemicals, heavy metals, and sediment. WASP is a flexible framework for modeling hydrodynamics, conservative mass transport, eutrophication-dissolved oxygen kinetics, and toxic chemical-sediment dynamics.

Experience   The original version of WASP dates back to 1981. The various versions of the model have been applied throughout the United States.

(CE-QUAL-RIV1) A Dynamic One-Dimensional Water Quality Model for Streams

Contact   Water Quality and Contaminant Modeling Branch
          Environmental Laboratory
          Waterways Experiment Station

U.S. Army Corps of Engineers
3909 Halls Ferry Road
Vicksburg, MS 39180-6199
(601)634-3785

Model Availability   CE-QUAL-RIV1 and documentation can be obtained by contacting the Waterways Experiment Station at the above address and telephone number. The computer program and documentation can be copied without restriction.

Documentation   The model is documented by "CE-QUAL-RIV1: A Dynamic, One-Dimensional (Longitudinal) Water Quality Model for Streams: User's Manual," Instruction Report E-90-1, U.S. Army Engineer Watersway Experiment Station, Vicksburg, Mississippi, November 1990.

Computer Configuration   An executable version of the model is available for MS-DOS-based microcomputers. The FORTRAN77 programs have been compiled and executed on various other computer systems as well.

Capabilities   CE-QUAL-RIV1 is a one-dimensional (longitudinal) fully dynamic hydraulic flow and water quality simulation model intended for modeling highly unsteady streamflow conditions, such as that associated with peaking hydroelectric power tailwaters. The model also allows simulation of branched river systems with multiple control structures such as reregulation dams and navigation locks and dams. The model has two parts, hydrodynamics and water quality. Output from the hydrodynamic model is used to drive the water quality model. The hydrodynamics are based on an implicit numerical solution of the St. Venant equations. The water quality constituents which can be modeled include temperature, dissolved oxygen, carbonaceous biochemical oxygen demand, organic nitrogen, ammonia nitrogen, nitrate nitrogen, orthophosphate phosphorus, coliform bacteria, dissolved iron, and dissolved manganese. The effects of algae and macrophytes can also be included.

Experience   The model was originally developed at Ohio State University for the U.S. Environmental Protection Agency for predicting water quality associated with stormwater runoff. The model was revised during the 1980s by Ohio State University and the USACE Waterways Experiment Station (WES). The current version has been tested and applied in several studies at WES.

(CE-QUAL-R1) A Numerical One-Dimensional Model of Reservoir Water Quality

Contact   Water Quality and Contaminant Modeling Branch
Environmental Laboratory

U.S. Army Engineer Waterways Experiment Station
3909 Halls Ferry Road
Vicksburg, MS 39180-6199
(601)634-3785

Model Availability CE-QUAL-R1 and documentation can be obtained by contacting the Waterways Experiment Station at the above address and telephone number. The computer program and documentation can be copied without restriction.

Documentation CE-QUAL-R1 is documented by "CE-QUAL-R1: A Numerical One-Dimensional Model of Reservoir Water Quality, User's Manual," Instruction Report E-82-1, Environmental Laboratory, U.S. Army Engineer Watersway Experiment Station, Vicksburg, Mississippi, July 1986.

Computer Configuration An executable version of the model is available for MS-DOS-based microcomputers. The FORTRAN77 programs have been compiled and executed on various other computer systems as well.

Capabilities CE-QUAL-R1 simulates the vertical distribution of thermal energy and chemical and biological materials in a reservoir through time. The model is used to study water quality problems and the effects of reservoir operations on water quality. A reservoir is conceptualized as a vertical sequence of horizontal layers with thermal energy and materials uniformly distributed in each layer. The distribution of inflows among the horizontal layers is based on density differences. Vertical transport of thermal energy and materials occurs through entrainment and turbulent diffusion. The interactions of numerous biological and chemical factors are reflected in the model. The model simulates the dynamics of 27 water quality variables, computing both in-pool and downstream release magnitudes. Eleven other variables are included which represent materials in the sediments. Reservoir outflows may occur in the model according to a specified schedule of port releases. Alternatively, the model may select port releases based on user specification of total release and desired release temperatures. Water quality problems that can be addressed include prediction and analysis of thermal stratification, anoxic conditions, algal blooms, and growth of algae and macrophytes; location of selective withdrawal ports required to meet a downstream temperature objective; and analysis of the effects of storm events, upstream land use changes, or reservoir operational changes on in-pool and release water quality.

Experience CE-QUAL-R1 has been applied in several studies conducted by the Waterways Experiment Station.

(CE-QUAL-W2) A Numerical Two-Dimensional, Laterally Averaged Model of Hydrodynamics and Water Quality

Contact   Water Quality and Contaminant Modeling Branch
Environmental Laboratory
Waterways Experiment Station
U.S. Army Corps of Engineers
3909 Halls Ferry Road
Vicksburg, MS 39180-6199
(601)634-3785

Model Availability   CE-QUAL-W2 and documentation can be obtained by contacting the Waterways Experiment Station at the above address and telephone number. The computer program and documentation can be copied without restriction.

Documentation   CE-QUAL-W2 is documented by "CE-QUAL-W2: A Numerical Two-Dimensional, Laterally Averaged Model of Hydrodynamics and Water Quality; User's Manual," Instruction Report E-86-5, Environmental Laboratory, U.S. Army Engineer Waterways Experiment Station, Vicksburg, MS, August 1986.

Computer Configuration   An executable version of the model is available for MS-DOS-based microcomputers. The FORTRAN77 programs have been compiled and executed on various other computer systems as well.

Capabilities   CE-QUAL-W2 was developed for reservoirs but can also be applied to rivers and estuaries. The two-dimensional model simulates the vertical and longitudinal distributions of thermal energy and selected biological and chemical materials in a water body through time. The model provides capabilities for assessing the impact of reservoir design and operations on the water quality variables. The model determines in-pool water volumes, surface elevations, densities, vertical and longitudinal velocities, temperatures, and constituent concentrations as well as downstream release concentrations. The unsteady flow hydrodynamic model handles variable density effects on the flow field. The water quality model simulates the dynamics of up to 20 constituents in addition to temperatures and circulation patterns. The model simulates the interaction of physical factors (such as flow and temperature), chemical factors (such as nutrients), and an algal assemblage. The constituents are arranged in four levels of optional modeling complexity, permitting flexibility in model application. The first level includes materials that are conservative and noninteractive. The second level includes the interactive dynamics of oxygen-phytoplankton-nutrients. The third level allows simulation of pH and carbonate species. The fourth level allows simulation of total iron.

Experience CE-QUAL-W2 has been applied in several studies conducted by the Waterways Experiment Station.

(HEC-5Q) Simulation of Flood Control and Conservation Systems (Water Quality Version)

Contact   Hydrologic Engineering Center
U.S. Army Corps of Engineers
609 Second Street
Davis, CA 95616
(916)756-1104

Model Availability   Separate versions of HEC-5, with and without water quality analysis capabilities, are available from the Hydrologic Engineering Center.

Documentation   The HEC-5Q water quality features are documented by an appendix to the HEC-5 user's manual: "HEC-5 Simulation of Flood Control and Conservation Systems, Appendix on Water Quality Analysis," USACE Hydrologic Engineering Center, Draft, September 1986. A training document and several papers on specific applications of the water quality model are also available from the HEC.

Computer Configuration   The water quality model HEC-5Q is an expanded version of the quantity-only HEC-5 with subroutines added to provide water quality analysis capabilities. The FORTRAN77 program was originally developed on mainframe computers but has been recently compiled to run on MS-DOS-based microcomputers.

Capabilities   The HEC-5 flow simulation capabilities are outlined separately herein under the category of Reservoir/River System Operation Models. The flows computed by the flow simulation submodel for multiple-reservoir and nonreservoir control points are input to the water quality submodel. The water quality simulation module computes the vertical distribution of temperature and other constituents in the reservoirs and the water quality in the associated downstream reaches. The model also determines the gate openings for reservoir selective withdrawal structures to meet user-specified water quality objectives at downstream control points. If the downstream quality objectives cannot be satisfied by selective withdrawal, the model will determine if the objectives can be satisfied by an increase in flow amounts. The water quality simulation can be used in three alternative modes: calibration, annual simulation, and long-term simulation. Two alternative groups of water quality constituents can be simulated. The first option includes water temperature, up to three conservative constituents, up to three nonconservative constituents, and dissolved oxygen. The other option includes water temperature, total dissolved solids, nitrate nitrogen, phosphate phosphorus, phytoplankton, carboneous BOD, ammonia nitrogen, and dissolved oxygen.

Experience   Although the quantity-only version of HEC-5 has been applied widely, the water quality version has been applied in only a relatively few studies.

(WQRRS) Water Quality for River-Reservoir Systems

Contact   Hydrologic Engineering Center
U.S. Army Corps of Engineers
609 Second Street
Davis, CA 95616
(916)756-1104

Model Availability   Hydrologic Engineering Center (HEC) programs, including WQRRS, are available through the National Technical Information Service (NTIS) and various other distributors. Federal agencies can obtain programs directly from the HEC. An HEC computer program catalog and list of distributors are available, free of charge, by contacting the HEC. The NTIS and private distributors charge a nominal handling fee. The programs and documentation can be copied without restriction.

Documentation   WQRRS is documented by "Water Quality for River-Reservoir Systems User's Manual (WQRRS)," USACE Hydrologic Engineering Center, October 1978 (revised February 1985).

Computer Configuration   The FORTRAN programs were originally developed on mainframe computers. Compiled versions are also available for MS-DOS-compatible microcomputer systems.

Capabilities   The WQRRS package consists of the programs SHP, WQRRSQ, and WQRRSR, which interface with each other. The Stream Hydraulics Package (SHP) and Stream Water Quality (WQRRSQ) programs simulate flow and quality conditions for stream networks, which can include branching channels and islands. The Reservoir Water Quality (WQRRSR) program is a one-dimensional model used to evaluate the vertical stratification of physical, chemical, and biological parameters in a reservoir. The SHP provides a range of optional methods for computing discharges, velocities, and depths as a function of time and location in a stream system. The hydraulic computations can be performed optionally using input stage-discharge relationships, hydrologic routing, kinematic routing, steady flow equations, or the full unsteady flow St. Venant equations. The WQRRSR and WQRRSQ programs provide capabilities for analyzing up to 18 constituents, including chemical and physical constituents (dissolved oxygen, total dissolved solids), nutrients (phosphate, ammonia, nitrite, and nitrate), carbon budget (alkalinity, total carbon), biological constituents (two types of phytoplankton, benthic algae, zooplankton, benthic animals, three types of fish), organic constituents (detritus, organic sediment), and coliform bacteria.

Experience   WQRRS is an advanced complex model which is relatively difficult to use. Consequently, it has not been applied widely. WQRRS has been applied in several Corps of Engineers studies.

RESERVOIR/RIVER SYSTEM OPERATION MODELS (CHAPTER 10)

(HEC-5) Simulation of Flood Control and Conservation Systems

Contact   Hydrologic Engineering Center
U.S. Army Corps of Engineers
609 Second Street
Davis, CA 95616
(916)756-1104

Model Availability   Hydrologic Engineering Center (HEC) programs, including HEC-5, are available through the National Technical Information Service (NTIS) and various other distributors. Federal agencies can obtain programs directly from the HEC. An HEC computer program catalog and list of distributors are available, free of charge, by contacting the HEC. The NTIS and private distributors charge a nominal handling fee. The programs and documentation can be copied without restriction.

Documentation   An HEC-5 user's manual and various other publications are available from the HEC, including training documents covering various features of the model and reports and papers documenting specific studies. A free publications catalog is available from the HEC upon request. HEC publications can be ordered directly from the HEC using the price list and order form provided in the catalog. Charges are waived for federal agencies. HEC-5 is also included in the series of short courses taught by the HEC.

Computer Configuration   HEC-5 was originally developed on mainframe computers and later compiled to run on MS-DOS-based microcomputers. The model is coded in FORTRAN77. The HEC-5 microcomputer software package includes several utility programs, linked by a menu structure, which facilitate developing input files and analyzing output. The COED editor and HEC-DSS (Data Storage System) are often used with HEC-5. HEC-5 is a large program with significant memory and disk storage requirements which vary between alternative versions of the model.

Capabilities   HEC-5 simulates multiple-purpose, multiple-reservoir systems on essentially any stream tributary configuration using a variable computational time interval. The model makes release decisions to empty flood control pools and to meet user-specified diversion, instream flow, and hydroelectric energy targets, based on computed reservoir storage levels and flows at downstream locations. Seasonal rule curves and buffer zones can be specified. Multiple-reservoir release decisions are based on balancing the percent depletion in user-specified storage zones. Several alternative hydrologic flood routing methods are available. Various optional analysis capabilities are provided, including computation of firm yields for diversions, instream flows, or hydroelectric energy and computation of expected annual flood damages. A water quality version of HEC-5 is also available.

Experience   HEC-5 has been used in numerous studies conducted by the HEC, other USACE offices, and other federal and nonfederal entities.

(IRIS) Interactive River System Simulation Model

Contact   Daniel P. Loucks
Civil & Environmental Engineering
Hollister Hall, Cornell University
Ithaca, NY 14853-3501
(607)255-4896

M. R. Taylor
Resources Planning Associates, Inc
Langmuir Building, Suite 231
Cornell Business & Technology
    Park
Ithaca, NY 14850
(607)257-4305

K. A. Salewicz
International Institute for
    Applied Systems Analysis
A-2361 Laxenburg
Austria
(043)(2236)715210

Model Availability The International Institute for Applied Systems Analysis and Resources Planning Associates, Inc. distribute the software and associated manuals. A nominal handling fee is charged. The user can copy the program and manuals without restriction.

Documentation Two IRIS manuals, (1) general introduction and description and (2) user's manual, are distributed with the program.

Computer Configuration The interactive, menu-driven model makes extensive use of color graphics. The program is written in FORTRAN77 using the CAPLIB graphics toolkit developed by Resources Planning Associates, Inc. Compiled versions of the model are available for MS-DOS-based microcomputers and VAX minicomputers with VMS.

Capabilities IRIS simulates water supply storage and conveyance systems of any normal branching configuration for given operating rules and streamflow sequences, using a user-specified time step. The model also includes hydro-electric power and water quality features. The configuration of the system is specified by "drawing in" nodes (reservoirs, inflow sites, junctions, and other key locations) and interconnecting links (river reaches, canals, and pipelines). System operating rules include (1) reservoir releases specified as a function of storage and season of the year; (2) allocation functions for multiple links from the same node; and (3) storage distribution targets for reservoirs operating as a group. Model output includes time series plots of flows, storages, energy generated, and water quality parameters at any node or link in the reservoir/river system and probability distribution displays of magnitude and duration of shortages or failure events.

Experience IRIS has been used in a number of studies.

(TAMUWRAP) Water Rights Analysis Package

Contact Dr. Ralph A. Wurbs
Civil Engineering Department
Texas A&M University
College Station, TX 77843
(409)845-3079

Texas Water Resources Institute
Texas A&M University System
College Station, TX 77843
(409)845-1851

Model Availability TAMUWRAP can be obtained by contacting either Ralph Wurbs or the Texas Water Resources Institute.

Documentation The model is documented by Wurbs, Dunn, and Walls, "Water Rights Analysis Package (TAMUWRAP) Model Description and User's Manual," Technical Report 146, Texas Water Resources Institute, March 1993.

Computer Configuration An executable version of the model is available for MS-DOS-based microcomputers. The FORTRAN77 programs have also been compiled and executed on VAX minicomputers with VMS. The TAMUWRAP package includes the simulation model and a utility program for organizing and summarizing the simulation results.

Capabilities TAMUWRAP is designed for analyzing water management within a water rights permit system, with water demands being met on the basis of specified priorities. A user-specified set of water demands are met, as water availability allows, following specified operating rules for inputted sequences of streamflows and evaporation rates. A monthly time step is used. The model provides the capability to simulate a stream/reservoir/use system involving essentially any stream tributary configuration. Interbasin transfers and closed loops, such as pipelines carrying water upstream or between tributaries, can be included in the system. Hydroelectric power can also be included. Water use requirements and reservoir operating rules are specified in various optional formats. Selected multiple reservoirs can be operated in combination based on balancing the percent depletion in specified storage zones. As currently dimensioned, the system can contain up to 2,000 water rights, and each right can include both reservoir storage and/or a water demand target. Simulation results include diversions, shortages, hydroelectric energy generated, streamflow depletions, unappropriated streamflows, reservoir storages and releases, reservoir evaporation, and reliability statistics. A recent salinity version of the model includes capabilities for inputting salt loads and specifying maximum allowable salt concentrations as part of the diversion requirements.

Experience The recently developed TAMUWRAP has been applied in several river basin studies in Texas, which has a statewide prior appropriation surface water rights system.

(MODSIM) River Basin Network Simulation Model

Contact Dr. John W. Labadie
Department of Civil Engineering
Colorado State University
Fort Collins, CO 80523
(303)491-8596

Model Availability   MODSIM can be obtained by contacting Dr. Labadie at the address and telephone number cited above.

Documentation   The various versions of MODSIM have been documented by reports, papers, and manuals. Documentation can be obtained by contacting Dr. Labadie.

Computer Configuration   Versions of MODSIM are available for MS-DOS-based microcomputers and Unix-based workstations. The FORTRAN program has been run on several other machines as well. The model operates in an interactive, menu-driven environment.

Capabilities   MODSIM is a generalized river basin network simulation model for hydrologic and water rights analysis of complex water management systems. Water is allocated based on user-specified priorities and operating rules. The user assigns relative priorities for meeting diversion, instream flow, and storage targets as well as lower and upper bounds on flows and storages. The model computes values for all pertinent flows and storages. Hydroelectric power operations can be included in the simulation. A version of MODSIM also includes a stream-aquifer interaction model for analyzing conjunctive use of surface and ground water. A network flow programming problem is solved for each individual time interval. Monthly, weekly, or daily time intervals may be used. The out-of-kilter algorithm incorporated in earlier versions of MOD-SIM has recently been replaced with a more efficient algorithm based on a LaGrangian relaxation strategy. MODSIM output includes various optional tabular and graphical presentations of reservoir balances, flows, demands satisfied from surface and ground water, demand shortages, and energy generated.

Experience   Various versions of MODSIM have been applied by a number of water resources management organizations as well as by researchers at Colorado State University in a variety of studies involving reservoir/river systems in both the United States and other countries.

(HEC-PRM) Hydrologic Engineering Center Prescriptive Reservoir Model

Contact   Hydrologic Engineering Center
U.S. Army Corps of Engineers
609 Second Street
Davis, CA 95616
(916)756-1104

Model Availability   Hydrologic Engineering Center (HEC) programs are available through the National Technical Information Service (NTIS) and various other distributors. Federal agencies can obtain programs directly from the HEC. An HEC computer program catalog and list of distributors are available, free of charge, by contacting the HEC. The NTIS and private distributors charge a nominal handling fee. The programs and documentation can be copied without restriction. The recently developed HEC-PRM has not yet been distributed widely.

Documentation   Model development is documented by the following Hydrologic Engineering Center project reports: PR-15, "Missouri River System Analysis Model—Phase I," February 1991; PR-16, "Columbia River System Model—Phase I," October 1991; PR-17, "Missouri River Reservoir System Analysis Model: Phase II," January 1992; and PR-21, "Columbia River System Analysis Model—Phase II," December 1993. A user's manual and related materials are provided as appendixes in PR-17 cited above. A comprehensive user's manual will be developed to support public release of the model.

Computer Configuration   HEC-PRM is a FORTRAN77 program which has been compiled to run on microcomputers with MS-DOS. The model is used in combination with the HEC-DSS (Data Storage System), which provides input data preparation and output analysis capabilities.

Capabilities   HEC-PRM is a network flow programming model which incorporates an economic objective function. Operation of the reservoir/river system is driven by user-inputted, convex, cost-based, piece-wise linear penalty functions. Thus, the user must be able to express costs associated with various system purposes as a function of reservoir storage, instream flows, or diversions. Noneconomic components can also be included in the basically economic objective function. Operating rules are also reflected in the upper and lower bounds specified on flows, releases, and storage. For given sequences of inputted stream inflows, the model computes the instream flows, diversions, and storages for each month of the simulation period, which minimizes the objective function. The computations are performed for all months simultaneously. Improved network flow computational algorithms have been developed in conjunction with HEC-PRM.

Experience   HEC-PRM was developed in conjunction with studies of reservoir system operations in the Missouri and Columbia River Basins.

(RSS) River Simulation System

Contact   Dr. Jacquelyn F. Sullivan
Center for Advanced Decision Support for Water
  and Environmental Systems (CADSWES)
University of Colorado, Campus Box 421
Boulder, CO 80309-0421
(303)492-3972

Model Availability   The RSS can be obtained by contacting CADSWES. Initial model development by CADSWES was sponsored by the Bureau of Reclamation. The Tennessee Valley Authority and the Electric Power Research Institute have also recently supported additional work by CADSWES in developing and applying an object-oriented modeling framework to the TVA reservoir system.

Documentation   The October 1992 RSS documentation package includes a user's manual, technical reference manual, user tutorial, Colorado River database reference manual, and set of selected readings.

Computer Configuration   The interactive, graphics-based RSS runs on workstations using the Unix operating system. The software package integrates several programs to manage input data and analyze output, including S-plus, ARC/INFO, INGRES, and HYDAS and other commercial programs. The main reservoir/river system simulation model is written in an object-oriented C/C++ programming language.

Capabilities   RSS combines interactive computer graphics and database management with river/reservoir system simulation. The object-oriented structure provides flexibility from both user and programmer perspectives. The user develops a model of a particular river/reservoir system by combining selected objects. Preprogrammed instructions for handling data and performing computations are associated with each object. For example, the user might select a reservoir object, powerplant object, or diversion object to represent a component of the system, which results in the model performing certain computations associated with these objects. Input and output data are also defined by user-selected objects. The user defines reservoir system operating policies using English-like statements, following a specified format, which utilize preprogrammed functions. If sufficient flexibility is not provided for a particular application by the available RSS objects and statement functions, a programmer can readily modify the code to change existing objects and functions or add new ones.

Experience The recently developed RSS as well as the general concept of constructing reservoir/river system operation models within an object-oriented programming environment have been applied in several studies but are still in a developmental stage.

(CALIDAD) Object-Oriented River Basin Modeling Framework

Contact Water Management Section, D-5755
U.S. Bureau of Reclamation
Denver Federal Center
P.O. Box 25007
Denver, CO 80225
(303)236-4215

Model Availability The public domain CALIDAD was developed by and is maintained by the Bureau of Reclamation. The model, with documentation, is available by contacting the Water Management Section of the Bureau at the above address and telephone number.

Documentation CALIDAD is documented by a programmer's manual and the following user's manual: "The CALIDAD Framework, User's Manual, Version 1.2," U.S. Bureau of Reclamation, Denver, Colorado, May 1994.

Computer Configuration The interactive, graphics-based CALIDAD runs on workstations using the Unix operating system. The object-oriented modeling system was developed using the C programming language and one of its object-oriented extensions, Objective-C. The graphical interface was developed in C using Motif and the X Intrinsics Libraries.

Capabilities CALIDAD simulates the movement of water through a reservoir/river basin system and determines the set of diversions and reservoir releases which best meets the institutional constraints and management objectives. Simulations are performed using a monthly computational time step. The user builds a model for a specific river basin application by using objects which represent features such as inflows, reservoirs, diversions, hydropower plants, and irrigation or municipal water demand sites. CALIDAD has a palette of available objects from which to choose. Additional objects can be programmed and added to the library as needed. Both computational algorithms and data requirements are associated with each object. The physical parameters of the river basin features, such as reservoir storage characteristics and monthly streamflows, may be entered as object data. Institutional constraints and management objectives, called rules in the model, are also considered as data and entered through a separate rules editor. CALIDAD handles the management and institutional constraints using a heuristic technique called tabu search to determine permissible diversion and reservoir releases. If the system is overconstrained, the tabu search selects a release schedule using weighting factors provided by the user.

Experience   The recently developed CALIDAD modeling system is still being tested and refined. The modeling package is being applied to several reservoir/river systems, including the Central Valley Project in California.

SOFTWARE INDEX

A&M Watershed Model; Texas A&M University; 113, 121, 198

ACCESS, database management; Microsoft Corp.; 51

ACRES Model, reservoir system analysis; Acres Consulting Services; 173

AGDAM, Agricultural Flood Damage Analysis; Hydrologic Engineering Center; 22

AGNPS, Agricultural Non-Point Source Model; Agricultural Research Service; 120

AGU-10, package of ground water flow and transport models based on American Geophysical Union Water Resources Monograph 10; International Ground Water Modeling Center; 96

AL-V, surface water allocation; Texas Water Development Board; 172

ANNIE, hydrologic analysis and data management; U.S. Geological Survey; 53

ANNIE-IDE, Interactive Development Environment; EPA Center for Exposure Assessment Modeling; 26, 53

AQUIFEM-1, ground water flow; Massachusetts Institute of Technology; 93

ARC/INFO, geographic information system; Environmental Systems Research Institute; 56

ARMOS, areal migration of free phase light hydrocarbon and recovery system design; Environmental Systems and Technologies; 96

ASHDRAIN, Design of Inlets and Drainage Networks; McTrans, Ashoke Kachroo; 31

ASM, ground water flow and transport; Kinzelbach and Rausch; 96

AT123D, analytical ground water solute transport; Oak Ridge National Laboratory; 96

AutoCAD, computer-aided drafting; Autodesk, Inc.; 59

Axum, graphics program; TriMetriz, Inc.; 61

BALANCE, ground water; International Ground Water Modeling Center; 96

BEAVERSOFT, ground water package; Bear and Verruijt; 93, 96

BETTER, Box Exchange Transport Temperature and Ecology of Reservoirs; Tennessee Valley Authority; 147

BIOPLUME II, Two-Dimensional Contaminant Transport Under the Influence of Oxygen Limited Biodegradation in Ground Water; EPA Center for Subsurface Modeling Support; 28, 96

BMDP/PC, statistical analysis; BMDP Statistical Software, Inc.; 48

BRASS, Basin Runoff and Streamflow Simulation; USACE Savannah District, Colon and McMahon; 166

CALIDAD, river basin simulation; Bureau of Reclamation; 175, 176, 225

CANVAS, transport and fate of viruses in ground water; Park, Blandford, Wu, and Huyakorn; 96

CAPZONE, ground water flow; Bair, Springer, and Roadcap; 96

CATTI, interpretation of tracer test data; Sauty and Kinzelbach, 96

CE-QUAL-R1, reservoir water quality; Waterways Experiment Station; 147, 149, 213

CE-QUAL-RIV1, dynamic stream water quality; Waterways Experiment Station; 148, 149, 215

CE-QUAL-W2, reservoir water quality; Waterways Experiment Station; 148, 149, 215

CFITIM, estimation of ground water solute transport parameters; USDA Salinity Laboratory; 96

CHEMFLO, One-Dimensional Water and Chemical Movement in Unsaturated Soils; EPA Center for Subsurface Modeling Support; 28, 96

CLC Database, Coordinated List of Chemicals Data Base; EPA Center for Exposure Assessment Modeling; 26

COBIAA, Consequences of Bioaccumulation in Aquatic Animals; Waterways Experiment Station; 44

CodeH2, Expert System for HEC-2; McTrans, Hydraulic Enhancements; 31

COED, Corps of Engineers Editor; Hydrologic Engineering Center; 22, 30

CONMIG, ground water contaminant migration; Walton; 93

COORS, Computation of Reservoir Stratification; Tennessee Valley Authority; 148

CorelDRAW, graphics; Corel Corp.; 60

CORMIX, Cornell Mixing Zone Expert System; EPA Center for Exposure Assessment Modeling; 26, 44

COVAR, generation of two-dimensional fields of autocorrelated parameters; International Ground Water Modeling Center, Williams and El-Kadi; 96

CRAM, Central Resource Allocation Model; WBLA, Inc., Brendecke, DeOreo, Payton, Rozaklis; 173

CREAMS, Chemicals, Runoff, and Erosion from Agricultural Management Systems; Agricultural Research Service, 119

Cricket Graph, graphics; Computer Associates International, Inc.; 61

CRSS, Colorado River Simulation System; Bureau of Reclamation; 158

CSUDP, Dynamic Programming; Colorado State University, Labadie; 48, 162

CXTFIT, estimation of ground water solute transport parameters; Virginia Polytechnical Institute; Parker and van Genuchten; 96

CYBERNET, pipe network analysis; Haestad Methods; 84

DABRO, Drainage Basin Runoff Model; McTrans, Golding; 31

DAMBRK, Dam-Break Flood Forecasting; National Weather Service; 131, 206

DAMP, Drainage Analysis and Modeling Programs; McTrans, Federal Highway Administration; 31

DBAPE, Data Base Analyzer and Parameter Estimator; EPA Center for Exposure Assessment Modeling; 26

dBASE, database management; Borland International, Inc.; 51

DHM, Diffusion Hydrodynamic Model; U.S. Geological Survey; 134

DOSAG, dissolved oxygen sag; Texas Water Development Board; 142

DR3M, watershed model; U.S. Geological Survey; 114

DREAM, analytical ground water flow; Bonn and Rounds; 93

DSS, Dam Site Selector; Purdue University, Engel and Beasley; 45

DSSMATH, Mathematical Utilities for DSS Data; Hydrologic Engineering Center; 22

DWOPER, Dynamic Wave Operational Model; National Weather Service; 131, 206

DWRSIM, reservoir/river system analysis; California Department of Water Resources; 174

DYNHYD5, Hydrodynamic Model; EPA Center for Exposure Assessment Modeling; 26

DYRESM, reservoir water quality; Imberger, Patterson, Herbert, Loh; 146

EASy, Engineering Analysis System; McTrans, Adges Software; 31

EasyPlot, graphics; Spiral Software; 61

Electronic Bulletin Board System; EPA Center for Exposure Assessment Modeling; 27

EML/IMES, Exposure Models Library and Integrated Model Evaluation System; Environmental Protection Agency; 27

ENDOW, Environment Design of Waterways; Waterways Experiment Station; 44

EPA-VHS, analytical ground water solute transport; International Ground Water Modeling Center; 96

EPIC, Erosion-Productivity Impact Calculator; Agricultural Research Service; 120

EUTRO, pollutants in riverine systems; Environmental Protection Agency; 144

EXAMSII, Exposure Analysis Modeling System II; EPA Center for Exposure Assessment Modeling; 26

EXCEL, spreadsheet; Microsoft Corporation; 35, 41

EXTEND, simulation modeling environment; Imagine That, Inc.; 38, 41

EXTRAN, dynamic flow routing; Environmental Protection Agency; 118, 131

FESWMS, Finite Element Surface Water Modeling System; McTrans, Federal Highway Administration, U.S. Geological Survey; 31, 133, 136, 208

FGETS, Food and Gill Exchange of Toxic Substances; EPA Center for Exposure Assessment Modeling; 26

FHAR, Flood Hydrograph and Routing; Bureau of Reclamation; 114

FLDWAV, Flood Wave; National Weather Service; 131, 136, 205

FLOWMASTER; McTrans, Haestad Methods; 31

FT, ground water; International Ground Water Modeling Center; 96

FTWORK, three-dimensional ground water flow and solute transport; GeoTrans, Inc.; 101

GAMS, General Algebraic Modeling System; Scientific Press; 47

GCSOLAR, Green Cross Solar Program; EPA Center for Exposure Assessment Modeling; 26

GEDA, Geometric Elements from Cross Section Coordinates; Hydrologic Engineering Center; 146

Generic CADD, Computer Aided Drafting and Design; Autodesk, Inc.; 60

GEO-EAS, geostatistical analyses; U.S. Environmental Protection Agency, Englund; 50, 96

GEOPACK, Geostatistics for Waste Management; EPA Center for Subsurface Modeling Support; 28, 50, 93, 96

GLEAMS, Groundwater Loading Effects of Agricultural Management Systems; Agricultural Research Service; 120

Grafit, graphics; Erithacus Software, Ltd.; 61

Graftool, graphics; 3-D Visions Corp.; 61

Grapher, graphics; Golden Software, Inc.; 61

GRAPHER, x-y plots; Golden Software, Inc.; 61

GRASS, Geographical Resources Analysis Support System; USACE Construction Engineering Research Laboratory; 57

GWFL3D, ground water flow; Walton; 93, 98

GWFLOW, ground water flow; International Ground Water Modeling Center; 96

GWGRAF, ground water graphics; Walton; 93

GWTR3D, ground water solute transport; Walton; 93, 98

Harvard Graphics; Software Publishing Corp.; 60

HEATX, Heat Exchange Program; Hydrologic Engineering Center; 22, 146

HEC-1, Flood Hydrograph Package; Hydrologic Engineering Center; 22, 30, 111, 121, 195

HEC-1F, Modified HEC-1 for Real-Time Water Control Systems; Hydrologic Engineering Center; 22

HEC-2, Water Surface Profiles; Hydrologic Engineering Center; 22, 30, 128, 136, 203

HEC-3, Reservoir System Analysis for Conservation; Hydrologic Engineering Center; 164, 165

HEC-4, Monthly Streamflow Simulation; Hydrologic Engineering Center; 22, 49

HEC-5, Simulation of Flood Control and Conservation Systems; Hydrologic Engineering Center; 22, 30, 165, 176, 218

HEC-5Q, Simulation of Flood Control and Conservation Systems with Water Quality Analysis; Hydrologic Engineering Center; 22, 145, 149, 216

HEC-6, Scour and Disposition in Rivers and Reservoirs; Hydrologic Engineering Center; 22, 30, 134, 136, 209

HECDSS, Data Storage System; Hydrologic Engineering Center; 22, 52

HEC-FDA, Flood Damages Analysis Package; Hydrologic Engineering Center; 22

HEC-FFA, Flood Frequency Analysis; Hydrologic Engineering Center; 22, 49

HEC-LIB, HEC Subroutine Library; Hydrologic Engineering Center; 22

HEC-PRM, Prescriptive Reservoir Model; Hydrologic Engineering Center; 22, 171, 176, 222

HEC-RAS, River Analysis System; Hydrologic Engineering Center; 130

HEC-12, FHWA Hydraulic Engineering Circular 12 (Pavement Drainage); McTrans, SMF Engineering; 31

HEC2ENTRY, HEC-2 Input; McTrans, Wadsworth; 31

HECWRC, Flood Flow Frequency Analysis; Hydrologic Engineering Center; 30

HGP, Hydraulics Graphics Package; Hydrologic Engineering Center; 22

HLDPA, Hourly Load Distribution and Pondage Analysis; North Pacific Division; 167

HMR52, Probable Maximum Storm; Hydrologic Engineering Center; 22, 30

HSP, Hydrocomp Simulation Model; Hydrocomp, Inc.; 116

HSPF, Hydrological Simulation Program-Fortran; Environmental Protection Agency; 26, 116, 121, 201

HST3D, three-dimensional heat and solute transport in ground water; U.S. Geological Survey, Kipp; 96, 101

HUMUS, Hydrologic Unit Model for the United States; Agricultural Research Service; 119

HY-8, Culvert Analysis; McTrans, Federal Highway Administration; 31

HY-EDIT, HEC-1 and HEC-2 Edit Program; McTrans, GKY and Associates; 31

HY-TB, Hydraulic Toolbox; McTrans, Federal Highway Administration; 31

HYCOST, Small-Scale Hydroelectric Power Costs Estimates; Hydrologic Engineering Center; 22, 30

HYDGEN, Watershed Hydrographs; McTrans; 31

HYDPAR, Hydrologic Parameters; Hydrologic Engineering Center; 22

HYDRAIN, Highway Drainage; McTrans, Federal Highway Administration; 31

HydroCAD, Computer Aided Design for Hydrology and Hydraulics of Stormwater Runoff; McTrans, Applied Microcomputer Systems; 31

HYDROSIM, reservoir system simulation; Tennessee Valley Authority; 163

HYDRUS, ground water flow and solute transport, Kool and van Genuchten; 96

HYDUR, Hydropower Analysis Using Streamflow Duration Procedures; Hydrologic Engineering Center; 22

HYMO, Hydrologic Model; Agricultural Research Service; 112

HYSSR, Hydro System Season Regulation; USACE North Pacific Division; 167

HYSYS, Hydropower System Regulation Analysis; USACE North Pacific Division; 167

ILLUDAS, Illinois Urban Drainage Area Simulator; 114

INFIL, ponded infiltration, International Ground Water Modeling Center; 96

IRAS, Interactive River-Aquifer Systems; Cornell University and Resource Planning Associates, Loucks, French, Taylor; 168

IRIS, Interactive River System Simulation; Cornell University and International Institute of Applied Systems Analysis; Loucks; Loucks, Salewicz, Taylor; 168, 176, 219

IWR-MAIN, Municipal and Industrial Water Use Forecasting System Institute for Water Resources; 73, 183

JBD2D/3D, ground water flow; U.S. Geological Survey, Bredehoeft; 96

Kentucky Watershed Model; University of Kentucky; 116

KYPIPE2, pipe network analysis; University of Kentucky; 85, 185

LAKECO, lake water quality; NOAA Great Lakes Environmental Research Laboratory; 146

LARM, Laterally Averaged Reservoir Model; Waterways Experiment Station; 148

LAST, Applied Stochastic Techniques; Bureau of Reclamation; 49

LC50, LC50 Values Estimation Program; EPA Center for Exposure Assessment Modeling; 26

LCA, Least Cost Analysis; McTrans, American Concrete Pipe Association; 31

Lindo, mathematical programming; Lindo Systems, Inc.; 47

LOTUS 1–2–3, spreadsheet; Lotus Development Corp.; 35–37, 41

LP88, linear programming; Eastern Software Products, Inc.; 47

Mac Culvert, analysis and design of culverts; McTrans, Kniel; 31

MacStorm Sewer, analysis and design of sewers; McTrans, Kniel; 31

MAP, Monitoring Analysis Package; Golder Associates; 96

MathCAD, mathematics; MathSoft, Inc.; 45

Mathematica, mathematics; Wolfram Research, Inc.; 45

MATLAB, mathematics; The MathWorks, Inc.; 45

MICROFEM, ground water flow; Hemker and van Elburg; 96

MicroStation, CADD; Intergraph Corp.; 60

MILHY, Military Hydrology; Waterways Experiment Station; 113

MINLAKE, lake water quality; St. Anthony Falls Hydraulic Laboratory; 146

MINTEQA2, geochemical equilibrium speciation; Allison, Brown, and Novo-Gradac; 26, 96

MITCAT, catchment model; Massachusetts Institute of Technology; 111, 114

MITSIM, river basin simulation; Massachusetts Institute of Technology, Center for Advanced Decision Support for

Environmental and Water Resources Systems, Strzepek; 167

MLRP, Multiple Linear Regression Program; Hydrologic Engineering Center; 22, 30, 49

MNDOT.HYD, Box Culvert Analysis; McTrans, Minnesota Department of Transportation; 31

MOC, method of characteristics ground water solute transport; U.S. Geological Survey; 93, 96, 98, 102, 191

MOCDENSE, modified version of MOC; U.S. Geological Survey; 96

MODFLOW, modular three-dimensional finite-difference ground water flow model; U.S. Geological Survey; 93, 96, 97, 102, 188

MODSIM, river basin network simulation; Colorado State University, Labadie; 171, 176, 221

MOFAT, Two-Dimensional Finite Element Program for Multiphase Flow and Multicomponent Transport; EPA Center for Subsurface Modeling Support; 28

MONITOR-I, surface water storage and conveyance systems; Texas Water Development Board; 164

MOTRANS, multiphase flow and transport of multicomponent organic liquids; Environmental Systems and Technologies; 96

MOUSE, Modeling of Urban Sewers; University of Denmark; 114

MS CLEAN, reservoir water quality; 146

MS-DOS, Microsoft Disk Operating System; 10

MSCLEAN, aquatic ecosystem; Park, Collins, Leung, Boyden, Albanese, de-Cappariis, Forstner; 146

MT3D, Modular Three-Dimensional Transport Model; EPA Center for Subsurface Modeling Support; 28

MULTIMED, Multimedia Exposure Assessment Model; EPA Center for Exposure Assessment Modeling; 26

NETPATH, geochemical mass-balance reactions; U.S. Geological Survey, Plummer, Prestemon, and Parkhurst; 96

NWSDSS, Load NWS Data Tapes in DSS; Hydrologic Engineering Center; 22

OASIS, Parameter Estimation System for Aquifer Restoration Models; EPA Center for Subsurface Modeling Support; 28

Ohio Watershed Model, Ohio State University; 116

ONED, ground water transport; USDA Salinity Laboratory, van Genuchten; 96

ONESTEP, estimation of parameters for soil hydraulic properties; Virginia Polytechnical Institute, Kool, Parker, and van Genuchten; 96

OPROUT, Stream Routing Optimization by Negative Local Flows; Hydrologic Engineering Center; 22

OPTP/PTEST, optimal well discharge; Asian Institute of Technology, Paudyal and Gupta; 96

PARADOX, database management; Borland International, Inc.; 51

PAS, Preliminary Analysis System for Surface Profile Computations; Hydrologic Engineering Center; 22, 31

PAT, pathlines and travel times for ground water flow; International Groundwater Modeling Center, Kinzelbach and Rausch; 96

Penn State Urban Runoff Model, Pennsylvania State University; 114

PESTAN, Pesticide Analytical Model; EPA Center for Subsurface Modeling Support; 28,96

PESTRUN, pesticide runoff simulation model; Kansas Water Resources Research Institute, McCall and Lane; 96

PHREEQE, geochemical reaction; U.S. Geological Survey; 96

PHRQPITZ, geochemical calculations in brines and other electrolyte solutions; Plummer, Parkhurst, Flemming, and Dunkle; 96

PLASM, Prickett-Lonnquist Aquifer Simulation Model; Illinois State Water Survey; 93, 96, 97, 102, 189

PLUME2D, analytical ground water

solute transport, International Ground Water Modeling Center; 96

PREMOC, preprocessor for MOC ground water flow model; Srinivasan; 96

PRISM, Potomac River Interactive Simulation Model; Potomac River Basin Commission, Johns Hopkins University; 157

PRMS, Precipitation-Runoff Modeling System; U.S. Geological Survey; 114

ProPlot, graphics; Cogent Software; 61

PRZM2, Pesticide Root Zone Model; EPA Center for Exposure Assessment Modeling; 26

PUMP, reservoir water quality; Waterways Experiment Station; 44

PUMPTEST, estimation of transmissivity and storage coefficient from time-draw down pump test data; Beljin; 96

QUAL I&II, stream water quality; Texas Water Development Board, Environmental Protection Agency; 142

QUAL2E, Enhanced Stream Water Quality Model; EPA Center for Exposure Assessment Modeling; 26, 45, 142, 148, 210

QUAL2E-UNCAS, Enhanced Stream Water Quality Model-Uncertainty Analysis; EPA Center for Exposure Assessment Modeling; 143

QUATTRO PRO, spreadsheet; Borland International, Inc.; 35, 41

RADFLOW, radial flow toward a well; University of Birmingham, Rathod and Rushton; 96

RANDOM WALK, ground water solute transport; Illinois State Water Survey; 96, 102, 190

REGFRQ, Regional Frequency Computation; Hydrologic Engineering Center; 22, 49

RESOP-II, reservoir operation; Texas Water Development Board; 172

RESQUAL II, reservoir water quality; University of Minnesota; St. Anthony Falls Hydraulic Laboratory; 146

RESTEMP, reservoir temperature; Tennessee Valley Authority; 146

RESTMP, Reservoir Temperature Stratification; Hydrologic Engineering Center; 22

RETC, Retention Curve; EPA Center for Subsurface Modeling Support; 28, 96

REZES, reservoir analysis; Simonovic; 165

RITZ, Regulatory and Investigative Treatment Zone Model; EPA Center for Subsurface Modeling Support; 28, 96

River Forecast System; National Weather Service; 116

RIVMIX, Prediction of Transverse Mixing in Natural Streams; Canada Centre for Inland Waters; 143

RMA-2, two-dimensional free surface flows; USACE, Resource Management Associates; 22, 133

RSS, River Simulation System; Center for Advanced Decision Support for Environmental and Water Resources Systems; 174, 176, 224

RWH, ground water solute transport; International Ground Water Modeling Center; 96

SARAH, Surface Water Assessment Model for Back Calculating Reductions in Abiotic Hazardous Wastes; Environmental Protection Agency; 143

SAS/ETS, statistical analysis; SAS Institute, Inc.; 48

SCOUR, Scour at Bridges; McTrans, Federal Highway Administration; 31

SDS, graphics; Datanalysis; 61

SELECT, reservoir water quality; Tennessee Valley Authority; 146

SHE, Systeme Hydrologique Europeen; Danish Hydraulic Institute, United Kingdom Institute of Hydrology, SOGREAH; 114

SHP, Stream Hydraulics Package; Hydrologic Engineering Center; 131

Sigma Plot, graphics; Jandel Scientific; 61

SIM-V, reservoir/river system analysis; Texas Water Development Board; 172

SIMYLD, reservoir/river system simulation; Texas Water Development Board; 172

SLAM, Steady Layered Aquifer Model; Aral; 93

SlideWrite Plus, graphics; Advanced Graphics Software, Inc., 61

SOIL, estimation of soil hydraulic properties; International Ground Water Modeling Center, El-Kadi; 96

SOILPROP, estimate soil hydraulic properties and associated uncertainty from particle size distribution data; Environmental Systems and Technologies; 96

SOLUTE, ground water solute transport; HydroLink, Beljin; 96

Spigot, streamflow synthesis; Cornell University; Grygier and Stedinger; 49

SPSS/PC+, statistical analysis; SPSS, Inc.; 48

SSANAL, Storm Sewer Analysis and Design Utilizing Hydrographs; McTrans, Golding; 31

SSARR, Streamflow Synthesis and Reservoir Regulation; USACE North Pacific Division; 115, 121, 167, 199

SSHYD, Storm Sewer Analysis and Design; McTrans, Golding; 27.

Stanford Graphics; 3-D Visions Corps.; 60

StatGraphics, statistical analysis; Statistical Graphics Corp.; 48

STATS, Statistical Analysis of Time Series Data; Hydrologic Engineering Center; 22, 30, 49

STEADY, stream water quality; Waterways Experiment Station; 143

STELLA, Systems Thinking, Experiential Learning Laboratory, With Animation; High Performance Systems, Inc.; 38, 41

STF, Soil Transport and Fate Database 2.0 and Model Management System; EPA Center for Subsurface Modeling Support; 28

STORM, Storage, Treatment, Overflow, Runoff Model; Hydrologic Engineering Center; 22

SUMMERS, soil cleanup levels; International Ground Water Modeling Center; 96

SUPER, reservoir/river system simulation; USACE Southwestern Division; 166

SURFER, contour maps and three-dimensional surface plots; Golden Software, Inc.; 61

SUTRA, saturated-unsaturated ground water solute transport; U.S. Geological Survey; 93,96,100, 102, 193

SWACROP, Soil Water and Crop Production Model; Wesseling, Kabat, van den Brock, and Feddes; 96

SWANFLOW, ground water flow; GeoTrans, Faust and Rumbaugh; 96, 102

SWAT, Soil and Water Assessment Tool; Agricultural Research Service; 119

SWATER, Optimal Sewer Design Package; McTrans, Eaglin and Wanielista; 31

SWICHA, seawater intrusion in coastal aquifers; GeoTrans, Lester; 96

SWIFT, Sandia Waste Isolation, Flow, and Transport; Sandia National Laboratories; 101, 102, 194

SWIP, Waste Injection Program; U.S. Geological Survey; 101

SWITCH, HEC-2 and WSPRO Utility Program; McTrans, Florida Department of Transportation; 31

SWM-IV, Stanford Watershed Model; Stanford University, Crawford and Linsley; 115

SWMM, Storm Water Management Model; Environmental Protection Agency; 26, 45, 117, 121, 200

SWMS 2D, two-dimensional ground water flow and solute transport; Simunek, Vogel, and van Genuchten; 96

SWRRB and SWRRB-WQ, Simulator for Water Resources in Rural Basins; Agricultural Research Service; 118, 121, 202

TABS-2, Open Channel Flow and Sedimentation; Waterways Experiment Station; 135

TAMUWRAP, Water Rights Analysis Package; Texas A&M University; 168, 176, 220

Tech*Graph*Pad, graphics; Binary Engineering; 61

TETRA, estimation of velocity components from hydraulic head data; International Ground Water Modeling Center, Srinivasan and Beljin; 96

Texas Watershed Model, University of Texas; 116

TGUESS, estimation of transmissivity from specific capacity data; Wisconsin Geological Survey, Bradbury and Rothchild; 96

THCVFIT, estimation of transmissivity and storage coefficient from pumping tests; van der Heijde; 96

THEISFIT, estimation of transmissivity and storage coefficient from test data, International Groundwater Modeling Center, McElwee and van der Heijde; 96

THERMS, Thermal Simulation of Lakes; Hydrologic Engineering Center; 22

THWELLS, ground water flow; International Groundwater Modeling Center, van der Heijde; 96

TIMELAG, estimation of hydraulic conductivity from time-lag tests; SRW Associates, Thompson; 96

TK! Solver, mathematics; Universal Technical Systems, Inc.; 45

TOUGH, ground water flow; Pruess; 102

TOXI, toxic pollutants in riverine systems; Environmental Protection Agency; 144

TR-20, Computer Program for Project Hydrology; Soil Conservation Service; 30, 112, 197

TR-55, Urban Hydrology for Small Watersheds; Soil Conservation Service; 30

TR20–88, Computer Program for Project Hydrology; McTrans, Soil Conservation Service; 31

TSSLEAK, estimation of parameters from pump test data; International Ground Water Modeling Center, McElwee and van der Heijde; 96

UHCOMP, Interactive Unit Hydrograph and Hydrograph Computation; Hydrologic Engineering Center; 22

UNET, Unsteady Flow through a Full Network of Open Channels; Hydrologic Engineering Center; Barkau; 22, 133, 136, 207

United Nations Ground Water Software; United Nations; 93

University of Cincinnati Urban Runoff Model; 114

UNWB-Loop, water distribution system analysis; World Bank; 84

USDAHL, watershed runoff; U.S. Department of Agriculture Hydrograph Laboratory; 116

USGMGT, Urban Stormwater Management Planning and Design; McTrans, Golding; 31

VaMP, Virginia Groundwater Mounding Program; McTrans, Expert Edge; 31

VARQ, estimation of aquifer parameters from pump test data; Kansas Geological Survey, Butt and McElwee; 96

VENTING, hydrocarbon recovery from unsaturated zone by vacuum extraction, Environmental Systems and Technologies; 96

VIRALT, transport and fate of viruses in ground water; HydroGeologic, Park, Blandford, and Huyakorn; 96

VisiCalc, spreadsheet; 35

VLEACH, Vadose Zone Leaching Model; EPA Center for Subsurface Modeling Support; 28, 96

WADISO, Water Distribution Simulation and Optimization; Waterways Experiment Station; 85, 187

WASP, Water Assignment Simulation Package; Kuczera and Diment; 173

WASP, Water Quality Analysis Simulation Program; Environmental Protection Agency; 26, 144, 149, 211

WATDSS, Load WATSTORE Data in DSS; USACE Hydrologic Engineering Center; 22

WATEQ4F, chemical equilibrium in natural waters; U.S. Geological Survey; 96

WATER, pipe network analysis; Bureau of Reclamation; 84

WATEXT, pipe network analysis; Bureau of Reclamation; 84

WATSTORE, Water Data Storage Retrieval system; U.S. Geological Survey; 22, 24

WEAP, Water Evaluation and Planning Model; Tellus Institute; 76, 184

WEATHER, weather data preprocessor for water quality programs; Hydrologic Engineering Center; 146

WELFLO, ground water flow; Walton; 93

WELFUN, Well Functions; Walton; 93

WELL, analysis of tracer test data; Gelhar; 96

WHPA, Wellhead Protection Area; Environmental Protection Agency; 28, 96, 99, 102, 192

WQRRS, Water Quality for River-Reservoir Systems; Hydrologic Engineering Center; 22, 30, 145, 149, 217

WQSTAT, Water Quality Statistics; Hydrologic Engineering Center; 22

WRMMS, reservoir water quality; Tennessee Valley Authority; 146

WSPRO GRAPH; McTrans, Softron; 31

WSPRO, Water Surface Profiles; U.S. Geological Survey and Federal Highway Administration; 31, 129, 136, 204

ZONEBUDGET, postprocessor for MODFLOW ground water model; U.S. Geological Survey, Harbaugh; 96

INDEX

Agricultural Research Service, 112, 118, 119
agricultural water requirements, 76
American National Standards Institute, 12
artificial intelligence, 42

BASIC programming language, 4, 12
Bureau of Reclamation, 19, 25, 175

calibration, 4, 6, 89, 101
California Department of Water Resources, 17, 174
Center for Microcomputers In Transportation, 29
central processor unit, 11
compiler, 11
computer-aided drafting and design, 59
computer modeling, 3
contaminant transport, 88–103, 116–122, 134–135, 139–149
C and C++ programming languages, 4, 12
Corps of Engineers (see U.S. Army Corps of Engineers)

database software, 35, 50
decision support systems, 14

demand forecasting, 68
desktop computers, 8
directories, software, 18, 34
disk operating system, 10
dynamic programming, 48, 161
dynamic routing, 130

electronic bulletin board, EPA CEAM, 26
Environmental Protection Agency (see U.S. Environmental Protection Agency)
equation solvers, 45
expert systems, 42
Exposure Models Library, EPA, 27

federal agencies, 20
Federal Computer Products Center, 29
Federal Highway Administration, 29, 129
FORTRAN programming language, 4, 11
    FORTRAN77, 11
    FORTRAN90, 12

generalized operational models, 5
geographic information systems, 54
geostatistics software, 49
graphical user interface, 9, 14

graphics software, 35, 60
ground water, 88–103

hardware, 7
high level languages, 11
highway drainage, 31
hydroelectric power, 22, 165, 167
Hydrologic Operational Multipurpose
    Subprogramme, 19

IBM-compatible microcomputers, 8
Illinois State Water Survey, 17, 97, 102
Integrated Model Evaluation System,
    EPA, 27
International Ground Water Modeling
    Center (IGWMC), 18, 28, 95
international organizations, 18

linear programming, 46, 161
LISP programming language, 43
local area networks, 9

machine language, 11
Macintosh microcomputers, 8
mainframe computers, 8
mathematical modeling environments, 45
mathematical programming, 46, 158–165
McTrans Center for Microcomputers in
    Transportation, 19
microcomputers, 8
microprocessor, 8
Microsoft Disk Operating System (MS-
    DOS), 8
minicomputers, 8
model applications, 36, 44, 58, 81, 88,
    107, 125, 139, 153
model categorization, 92, 109, 126, 141,
    155
modeling environments, 37, 40
modeling systems, 13
movable bed models, 134
MS-DOS, 10
multiple-tasking, 9
municipal and industrial water use fore-
    casting, 68

National Technical Information Service
    (NTIS), 19, 29

Federal Computer Products Center,
    29
National Weather Service (NWS), 19, 24,
    131
network flow programming, 169
NexGen, Hydrologic Engineering Center,
    112

object-oriented programming, 12, 174
object-oriented system simulation soft-
    ware, 37, 174
optimization, 46, 158–165
Office of Technology Assessment, U.S.
    Congress, 2
operational generalized models, 5
OS/2 operating system, 10

Pascal programming language, 12
PC-DOS, 10
Pentium microprocessor, 8
personal computers, 8
pipe network analysis, 81–86
post-processor programs, 14
precipitation-runoff models, 107–122
pre-processor programs, 14
programming languages, 11, 40
PROLOG programming language, 43
proprietary software, 13
public domain software, 13

QuickBASIC, 12

reservoirs, 139–149, 153–178
river hydraulics, 125–138

Sandia National Laboratories, 101
sediment transport, 134
Smalltalk programming language, 12
Soil Conservation Service, 19, 25, 112
source code, 11
spreadsheet software, 35
statistical analysis software, 48
systems software, 10
supercomputers, 8

temperature, 91, 140
Tennessee Valley Authority, 37
Texas Water Development Board, 17, 172
TURBOPascal, 12

two-dimensional models:
ground-water, 92
river hydraulics, 133
reservoir/river water quality, 147

United Nations, 18, 47, 93
U.S. Army Corps of Engineers
(USACE), 20
Construction Engineering Research
Laboratory, 57
Hydrologic Engineering Center
(HEC), 19, 20, 52, 111, 118, 128,
145
Institute for Water Resources (IWR),
19, 21, 70, 73
North Pacific Division, 115, 167
Waterways Experiment Station
(WES), 19, 23, 113, 144
U.S. Bureau of Reclamation, 19, 25, 175
U.S. Congress, Office of Technology As-
sessment, 2
U.S.D.A. Agricultural Research Service,
112, 118, 119
U.S. Environmental Protection Agency, 25
Center for Exposure Assessment
Modeling (CEAM), 19, 26, 116,
142, 144
Center for Subsurface Modeling Sup-
port (CSMoS), 19, 27, 95
Environmental Research Laboratory
in Athens, Georgia, 25

Office of Health and Environmental
Assessment, 27
Robert S. Kerr Environmental Research Laboratory, 27
U.S. Geological Survey (USGS), 23
Water Resources Division, 19, 23,
53, 94, 102, 114, 129
Water Resources Scientific Information Center, 24
U.S. Soil Conservation Service, 19, 25,
112
Unix, 10
unsteady flow models:
pipes, 83
ground water, 90
rivers, 130–135, 143–146

Visual Basic, 12

water conservation, 70
water distribution system analysis, 81–86
water management, 1
water quality, 88–103, 116–122, 134–
135, 139–149
watershed models, 107–122
water surface profiles, 128
WATSTORE, 24, 52
Windows, 10
workstations, 8
World Meteorological Organization, 19